SS PROFESSIONALS

ADVA
CIRO
TRAINING

complete guide to progressive planning and instructing

st edition

EBBIE LAWRENCE AND BOB HOPE

A & C BLACK · LONDON

Thanks to Fitness Professionals Ltd (www.fitpro.com) for supporting the Fitness Professionals series.

Note

It is always the responsibility of the individual to assess his or her own fitness capability before participating in any training activity. Whilst every effort has been made to ensure the content of this book is as technically accurate as possible, neither the author nor the publishers can accept responsibility for any injury or loss sustained as a result of the use of this material.

First published in 2008 by
A & C Black Publishers Ltd
38 Soho Square
London W1D 3HB
www.acblack.com

20th February 2009
540500002 263905

ISBN 978-1-408-10050-9

A CIP catalogue record for this book is available from the British Library.

The right of Debbie Lawrence and Bob Hope to be identified as authors of this work has been asserted by them in accordance with the Copyright Designs and Patents Act, 1988.

Typeset in Berthold Baskerville Regular by Palimpsest Book Production Limited, Grangemouth, Stirlingshire

Cover image © Grant Pritchard
Inside photography © Grant Pritchard, except pages 42 and 55 © Shutterstock
Illustrations © Jeff Edwards

Printed and bound in Spain by GraphyCems

This book is produced using paper from wood grown in managed, sustainable forests. It is natural, renewable and recyclable. The logging and manufacturing processes conform to the environmental regulations of the country of origin.

CONTENTS

ACKNOWLEDGEMENTS

The sense of pleasure I gain from writing far exceeds the very occasional sense of pain (writing blocks). I am eternally grateful that I have the opportunity to do what I love – write, teach and contribute to education and growth.

I have to admit that I am not a hardcore trainer! Most of the exercises within this book would be beyond my physical capacity and motivation. However, I have an admiration for people who push themselves to achieve physical self-actualisation.

Special appreciation for:

• The 'creative genius' of my co-author Richard (Bob) Hope. Firstly, for making the suggestion to write this book as a sequel to our original circuit book and secondly, for his ever flowing ideas on designing circuits and hardcore exercises. I definitely could not have written this book without you. Respect!
• The love, support, encouragement and positive energy from my partner and best friend Joe, who always makes me smile and laugh.
• The team of editors at A & C Black (Alex, Lucy, Robert, Ruth, Charlotte) for being a fabulous team to work with!

Debbie Lawrence (2007)

I would like to thank Debbie Lawrence, my co-writer and friend, for believing in the content material and agreeing to work with me on this project. Debbie, it would not have got off the ground without you and your huge exercise knowledge and powers of persuasion in the right circles. For this I am truly grateful to you, and so should be all readers of this book. Thank you.

To A&C Black for giving us the opportunity to co-write this, the second book based on circuit training.

To my mate Steve Plant at the Isle of Wight College who always has an ear open for me, to Lee Land from Plymouth YMCA and to Steve Platt at St Mellions in Cornwall – your help and support is always invaluable.

To my wife Jan, my greatest supporter, soul mate and best friend. Thanks for always being there for me.

Finally I would like to dedicate the book to my guardian angels:

John and Grace Hope
and
George Pettet

Richard Hope (2007)

INTRODUCTION

The aim of this book is to build on the information provided within the *Fitness Professionals: Circuit Training A Complete Guide to Planning and Instructing*, which was aimed at instructors qualified to Level 2 National Occupational Standards (NOS). Specifically, this book aims to:

- Develop instructor knowledge base in relation to each of the components of physical fitness to Level 3 of the NOS.
- Improve the understanding of the relationship between basic exercise and fitness knowledge and the components of fitness.
- Introduce different types of equipment that can be integrated into an advanced circuit training session.
- Provide exercise ideas with illustrations and descriptions, that focus on delivering more advanced circuit training programmes.
- Review safety considerations for delivering advanced circuit programmes/exercises.
- Introduce instructors to skill and learning theories that represent the knowledge required from a Level 3 instructor.

The information in this book represents knowledge required to meet the stated learning outcomes within the National Occupational Standards (NOS) and Common Units for Fitness Instruction (Level 2 and 3). This information can be applied in a number of different training contexts – exercise to music, gym-based exercise etc. The aim of this book is to discuss the concepts generally and then apply them to circuit training.

NOS Level 2 instructing exercise and fitness

Unit C35: deal with accidents and emergencies
Unit D416: evaluate coaching sessions and deal with personal coaching practice
Unit D417: support participants in developing and maintaining fitness
Unit D414: plan and prepare a group exercise with music session
Unit D415: instruct a group exercise with music session

NOS Level 3 instructing exercise and fitness

Unit D437: collect and analyse information to plan a progressive physical activity programme
Unit D438: plan, review and adapt a progressive physical activity programme
Unit D439: plan and instruct specific physical activity programmes (ETM)
Unit C313: provide motivation and support to clients during a progressive physical activity programme
Unit A318: manage, evaluate and improve your own performance in providing physical activity
Unit D444: integrate core stability and flexibility activities into a physical activity programme

NOS Level 3 core exercise and fitness knowledge

- Behaviour change
- Anatomy
- Functional kinesiology
- Energy systems
- Concepts and components of fitness

NB: copies of the Award Body Common Units and National Occupational Standards can be obtained from the Sector Skills Council (Skills Active) and/or the Register of Exercise Professionals (REPs).

Recommendation from the authors

As with all exercise programmes, there should be a focus on using correct exercise technique and posture! The equipment and exercises described in this book are specifically targeted at advanced exercisers who train frequently, are able to perform with safe exercise technique and have no specific medical conditions. This may include individuals who are serving within the armed forces, trained athletes and sport people.

To state what is hoped to be the obvious, these exercises are NOT recommended to be used when training beginners, people who are unused to training, children, older adults, special populations or people with specific medical conditions!

FITNESS GOALS

Physical fitness incorporates a number of components. These include cardiovascular fitness, muscular strength, muscular endurance, flexibility and motor fitness.

Physical fitness can be improved by performing an appropriate volume (frequency, intensity, time) of specific types of exercises and activities that are appropriately structured and planned.

The volume of exercise undertaken by individuals will be determined by:

- Specified training goals
- Current fitness level
- Current levels of activity and training
- Individual differences (age, body type etc)
- Any special consideration (medical conditions etc)

Chapter 1 reviews the principles and variables for progressive fitness training – progressive overload, frequency, intensity, time, recovery periods, reversibility, specificity, individual differences that affect performance and response to training.

Chapter 2 discusses cardiovascular training theory – the aims and benefits of cardiovascular (CV) training, training guidelines, methods of monitoring intensity, energy systems, progressive methods, different types of training.

Chapter 3 discusses flexibility training – the aims and benefits, training guidelines, factors that influence flexibility, different types of stretching, progressive methods.

Chapter 4 discusses muscular fitness training (strength and endurance) – aims and benefits, physiological changes, training guidelines, progressive methods, different training approaches.

Chapter 5 looks at motor skills training (power, speed, balance, agility etc) – with a brief introduction to the nervous system.

PRINCIPLES AND VARIABLES FOR PROGRESSIVE TRAINING

There are a number of principles and variables that need to be considered to ensure training programmes are both effective and safe. These include:

- Progressive overload (frequency, intensity, time)
- Recovery time
- Reversibility
- Specificity
- Individual differences (body type, age, gender etc)

This chapter introduces each of the variables and highlights specific considerations that need to be made when planning and designing fitness programmes. These considerations are explored in more detail in subsequent chapters.

Progressive overload

In order to progress, it is essential to make the body work a little bit harder than it has been used to working. This is achieved by increasing the volume of work (frequency, intensity, time) and also balancing training sessions with sufficient rest and recovery time.

The level of overload applied will be determined by the fitness goals of the individual and their current workload and fitness levels. A trained athlete will need a greater challenge to bring about improvements than an untrained, sedentary person, who may need very small incremental changes to bring about a training effect.

Rest and recovery

Rest and recovery time is essential to prevent over training. Factors that may influence the rest and recovery time needed may include the volume of exercise and also the fitness level of the individual. Periodised training programmes offer a method for structuring a workout plan to ensure sufficient volume of activity and rest. These are discussed in part 2.

For example:
An elite athlete may have a higher volume of work. They need sufficient rest time to enable the body to heal. An untrained person would have a lower volume of work, but their body may need longer to recover because it is unused to the specific training demands being made.

The specific component being trained may also influence the rest and recovery time needed. This is discussed in relation to frequency of training.

Frequency — how often we train

It is essential to train at an optimal frequency to achieve benefits from specific training regimes.

Optimal frequency of training, to improve specific components of fitness, is explored in this chapter.

For some components of fitness, training more frequently is beneficial. For example, stretching daily can provide greater improvements for flexibility than stretching on the minimum requirement of two to three days a week. However, for other components of fitness, rest time is needed between sessions to enable muscles to recover from the workout.

Muscular fitness training can be performed more frequently than the general guideline of two to three times a week. However, it is recommended that the same muscle groups are trained on non-consecutive days. Therefore, to train more frequently would require the use of a split training routine (training different muscles on different days) to enable sufficient rest and recovery time between sessions.

Cardiovascular fitness training can be performed more frequently. Most endurance athletes will often train four to five days a week, with some training more often. However, when training more frequently, the intensity of the workout would need to be modified during the weekly programme. For example, easy workouts rather than more intense workouts on some days.

Training muscular fitness and cardiovascular fitness too frequently and with insufficient rest between sessions can contribute to delayed onset muscle soreness (DOMS) and overtraining. Both can diminish the training benefits achieved.

Intensity – how hard we train

The intensity at which an individual will need to train will be determined by their starting point (existing fitness and skills level and current training regime) and specific training goals.

The range of potential variables to alter intensity include:

- Resistance
- Repetitions
- Range of motion
- Rest
- Sets
- Rate/speed

A key consideration would be how the specific variable can be manipulated and changed and how this will affect the training result. Each variable will bring about specific gains. For example, lifting heavier resistances for lower repetitions will bring about strength gains whereas lifting lighter resistances for higher repetitions will bring about endurance gains. Likewise, with cardiovascular training, working at different percentage of MHR will challenge different energy systems and this in turn will influence the time/duration for which the activity can be performed. Interval training requires short sets of higher intensity cardiovascular activities combined with active rest. This method is used for challenging the energy systems in a progressive way, which in turn enables the body to respond and adapt.

Time

The duration of training programmes is explored further in the following chapters. The timing of specific programmes will be determined by the component of fitness that is the focus of training and the specific fitness goals.

A key consideration when increasing the duration of training is the potential for this to affect adherence (American College of Sports Medicine: 2006). Increased duration may be less likely to affect the commitment of a dedicated athlete. However, it may be a factor that influences continued participation from a non-athlete or sedentary person.

> For example:
> A workout designed for general fitness may need less time (45 minutes to 1 hour) to train all components. Whereas, a more specific focus to meet a specific goal (a long distance run) may demand that a greater duration and focus in the session is placed on training for the specific goal - running for longer (increased time).

Reversibility – use it or lose it!

When training stops so too will the benefits gained stop. The muscles and other body systems need to be trained at an appropriate volume (frequency, intensity, time) to keep them functioning.

Not using the body systems through inactivity can lead to health problems (obesity, CHD etc). The Department of Health (2004, 2005) has proposed minimal activity guidelines for maintaining health.

- Frequency – 5 days a week
- Intensity – moderate
- Time – 30 minutes, which can be broken down and accumulated (3 x 10 minutes or 2 x 15 minute sessions)
- Type – daily activities performed with a little more vigour (walking, using stairs etc)

The more active and used to exercise a person is, the slower will be the rate of reversibility. Training programmes planned for athletes and sports people will need to include some down time (detraining), where hard training is reduced and the focus is on maintaining a base level of fitness. Periodised training programmes which plan for pre- and post-season training will make provision for the down time needed to enable the body to recover. This will be discussed further in part 2.

Specificity

Specific types of training programmes and exercises will induce specific physiological gains and benefits. Quite simply, this means cardiovascular training will improve cardiovascular fitness; it will not improve flexibility. Likewise, focusing on muscular strength training will not improve muscular endurance. The benefits and physiological gains achieved from training are specific to the component of fitness being trained.

Specific training modes will also bring about very specific and different training adaptations. Cardiovascular fitness will be improved by both running and swimming. However, this does not necessarily lead to runners being effective swimmers, nor does it lead to swimmers being effective runners. The body will adapt specifically to the demands made upon it – the specific adaptation to imposed demand (SAID) principle. Different activities will use different muscles, in different ways, with different energy demands and in the instance of running and swimming, different resistances (gravity/buoyancy).

Within general fitness programmes, an all round fitness is recommended with some cross training and variation of activity to minimise repetitive stress being applied to the body. However, individuals training for specific sporting and/or athletic events should pay greater attention to the specificity principle for their training. Different approaches to the structure of training programmes for specific goals will be reviewed in part 2 – periodisation and seasonal training. To reiterate the specificity principle – a marathon runner would need to run long distances within their programme. A channel swimmer would need to include long distance swims within their programme. A thrower would need to include some activity that mirrored the throwing action of their specific sport (javelin, shot putt, caber etc), a sprinter would need to sprint, a weight lifter would need to lift weights and a dancer would need to dance. This does not mean that other activities and fitness components are excluded. It means that consideration must be given to the specificity principle when planning programmes.

Some of the specificity factors will include:

- The components of fitness
- The sport/athletic event
- The energy system and muscle fibre type used most predominately
- The main prime movers recruited and joint actions
- The type of muscle contraction (concentric/ eccentric)
- The joint angle and range of movement for muscle work
- The speed of movement

Components of fitness

Each component of fitness has specific training requirements – frequency, intensity, time and type. The training requirements to improve flexibility are different to those for improving cardiovascular fitness, motor skills, muscular endurance and muscular strength.

Different exercises and activities will bring about different gains. Stretching the hamstrings will improve flexibility in the hamstrings, it will not improve strength of the hamstrings, nor will it improve flexibility in the triceps. Strengthening the pectorals will not improve strength of the quadriceps, nor will it develop cardiovascular fitness. Working on co-ordination to perform a dance sequence will not develop the co-ordination required for boxing. Each sport and activity has specific skills attached and will make demands on specific components of fitness.

The sport/athletic event

The training needs of a dancer will be different to the training needs of a boxer or a sprinter. Each athlete will need a good all round general fitness. They will also need to excel in specific areas depending on the demands of their sport. Within this, some alternative sporting activities may benefit performance in other sports. For example, footballers and rugby players will often not focus on flexibility training but if they were to take part in a dance class, where flexibility is a focus, their performance may be improved. Footballers have notoriously inflexible hamstrings and are susceptible to injury in this area. Improving their flexibility (through dance or other similar activities) may help reduce the risk of injury and enhance their performance during their game.

Energy system and muscle fibre type used

Different activities will use different energy systems and different muscle fibres. Marathon runners and long distance swimmers will require greater levels of aerobic fitness and will use predominately slow twitch muscle fibres. Sprinters and power lifters will require greater levels of anaerobic fitness and will use predominately fast twitch muscle fibres. Weight training with different resistances and repetition ranges will engage different energy systems and muscle fibres. Heavy lifting with lower repetitions will be predominately anaerobic (creatine phosphate system) and will use fast twitch muscle fibres. Lighter resistance training with repetition ranges exceeding 15–20 repetitions will be anaerobic (lactic acid) and most likely recruit intermediate fibres.

Working with very light resistances for much higher repetitions will use slow twitch muscle fibres and the aerobic energy system.

The main prime movers recruited and joint actions

Each activity and exercise will use specific muscles. Bicep curls will engage the biceps as the prime mover (main muscle working), sit-ups or curl-ups will engage the abdominal muscles as the prime mover, squats will engage the gluteals and quadriceps as the prime movers.

The main muscle groups used to perform throwing activities will be different to the main muscle groups used to perform jumping or leaping activities. Training programmes will need to consider the specific muscle actions used within the sporting activity.

To achieve a balanced muscular workout exercises for the whole body should be included. This can be within one training session (general fitness) or through a number of training sessions (split routine).

The positioning of the body in relation to gravitational forces will also influence the

For example:
Chest press action and bent over row action.

For both exercises the joint moving is the same and the joint action is the same (horizontal flexion and extension at shoulder joint and flexion and extension of elbow). However, the body position and thus the prime movers will be different. For the chest press the body is in a supine position (facing up to push the resistance) and for bent over rows the body is in a prone position (facing down to pull the resistance), thus the muscles being worked will be different. Chest press will work triceps, anterior deltoids and pectorals. Bent over row will work biceps, latissimus dorsi and trapezius.

muscles working and the type of muscle contraction (discussed below). Performing the same joint action but in a different direction against the gravitational forces will change the emphasis of muscle work.

The type of muscle contraction (concentric/eccentric) and gravity

Racquet sports such as squash demand a fair amount of leg lunge movements that require eccentric muscle work of the quadriceps and gluteals. Training to improve performance for squash would therefore demand some attention be paid to improve eccentric strength of the quadriceps and gluteals in the specific lunge position. The specificity principle requires consideration of the specific action and contraction of the muscles working. It also requires consideration to the joint angle and range of movement being performed (discussed next).

Activities performed in water require different types of muscle contraction to those based on land. This is because exercising in water reduces the influence of gravity on the body but adds the elements of buoyancy and floatation. Water also adds a number of different resistances to movement, which alter the muscle contraction. People interested in exploring the effects of water on movement are referred to the *Complete Guide to Exercise in Water* (Lawrence: 2004a).

The joint angle and range of movement for muscle work

Using the example of squash described previously, leg lunges demand eccentric muscle contraction of quadriceps and gluteals in a specific range of movement. In terms of specificity, this range of movement would need to be included in training. Performing seated leg extensions would provide eccentric muscle work for the quadriceps; however, the joint angle and range of movement would be different from that of a lunge. Lunges are a compound movement (more than one muscle involved) and closed chain kinetic pattern. Leg extensions are an isolated movement (one prime mover) and open chain kinetic pattern.

The speed of movement

The speed at which specific movements are performed within specific sporting activities will also need to be considered. If movements need to be performed quickly, then some attention should be paid to performing them at the speed required when training. For example, to perform well in a timed sit-up test, it is more effective to practice performing sit-ups quickly, rather than at a slower pace. If speed is the focus, then it is essential to train to develop speed.

The speed at which exercises are performed may also influence the type of muscle contraction. Performing exercises quickly during the lifting (concentric) phase of a movement and more slowly during the lowering (eccentric) phase of a movement, will place greater emphasis on the eccentric phase.

Individual differences

A further consideration for the specificity principle is that different individuals will respond differently to specific regimes and exercises. Some of the factors and differences among individuals will now be discussed.

There are numerous factors that will affect an individual's training potential. Each factor will need to be taken into consideration when designing and planning specific programmes.

Body type

There are three main body types (somatotypes) that contribute to differences in training potential. Generally speaking, most people share characteristics of each type, with a tendency towards either mesomorph and endomorph or mesomorph and ectomorph types. However, some individuals have higher characteristics of one specific body type. Table 1.1 describes characteristics of different body types.

Age

All body systems change throughout the lifespan. During formative years the body is growing and developing and consideration will need to be made to the effects of training on the developing body. The training guidelines for children and young adults are different to those recommended for adults. People interested in working with children should seek specialist training and are guided to specific texts (Harris: 1997).

During senior years, the changes to the body systems will reduce training potential. Benefits can still be obtained but the rate at which developments occur will be slightly slower. Age related changes generally begin to occur at 50 years old, and make their mark at around 65 years. Age related changes are listed in Table 1.2.

Regardless of age, an inactive lifestyle and little use of the muscles may contribute to the early onset of ageing and have negative effects on health (obesity, high blood pressure, low bone density etc). An active 70 or 80 year old may be in better shape and condition than someone who is inactive and 40. People interested in working with older adults should seek specialist training.

Table 1.1	Body types		
	Endomorphs	**Mesomorphs**	**Ectomorphs**
Shape	O	V	I
Individuals with body type	Winston Churchill Biggie Smalls Giant Haystacks	Charles Atlas LL Cool J Linford Christie	Zola Budd Snoop Dogg Paula Radcliffe
Illustration	 Endomorph	 Mesomorph	 Ectomorph
Characteristics	Thicker skeletal frame	Broader skeletal frame	Long, lean and narrow
	Shorter limbs/levers with tendency to lack of muscle tone. Rounder appearance Higher percentage of body fat. Lower muscle tone	Strong bones Wider shoulders Broad chest Narrower hips Dense muscularity	Skeletal frame Sunken chest Longer limbs/levers Lower levels of muscle and fat. Leaner appearance
Preferred activities	Endomorphs are well equipped to perform long distance swimming events. Their body fat assists floatation and will also help to keep them warm.	Mesomorphs have the physical potential (muscularity) to respond well to a variety of programmes. They often excel at sporting events and activities that require power and speed: sprinting, hurdling, throwing, jumping.	Ectomorphs are naturally equipped to perform longer distance running events. Their long levers, light body weight and streamlined narrow skeletal frame provides a comparatively lower resistance to movement which assists their performance in such activities.
Considerations	Endomorphs have a tendency to gain weight. Higher levels of body fat and low muscle tone make them less equipped to	Muscle bulk may restrict the potential for mesomorphs to develop flexibility. Muscularity will also add	Ectomorphs struggle to build muscle mass. The length of levers make it biomechanically harder for ectomorphs to lift a

Table 1.1	Body types (cont.)		
	perform higher impact and endurance activities. These activities would place additional stress to the joints. Endomorphs can build muscle size but this can exaggerate their natural shape and give them a larger appearance.	weight to skeletal frame, thus care should be taken with higher impact activities. The positive aspect is that their muscle strength provides resilience to impact stress.	resistance sufficient to build muscle bulk (hypertrophy) without placing unnecessary stress on their skeletal frame.

Gender

There are a number of anatomical and physiological differences between men and women that will influence their training potential. These should be considered when planning specific training programmes.

Skeletal frame

There are some structural differences in the shape of the male and female skeleton. The female pelvis is equipped for pregnancy and childbirth. It has wider and deeper shape. The male pelvis is longer and narrower. The width of the pelvis and the angle at which the femur is positioned in the hip socket is also different between men and women. The Q angle (quadriceps angle) of the female pelvis is greater and this gives women a biomechanical disadvantage when running or squatting. The greater Q angle leads to an inward rotation of the knees. This potentially may be a factor that influences differences in running speeds between men and women – women have to move their legs through a wider angle.

Body composition

Women generally have less muscularity and higher levels of body fat than men to equip them for pregnancy and childbirth. In terms of training, this would make it harder for women to gain muscle and lose body fat. Women may therefore need slightly longer to observe the effects of their training – definition of muscle for sports such as body building. A further consideration would be that for women to lower their body fat levels too low, would have detrimental effects of their health. Reducing body fat below recommended levels would reduce levels of the female hormone and may cause the cessation of the menstrual cycle. This in turn would contribute to lower levels of calcium and increased risk for osteoporosis.

Hormones

The most obvious difference between men and women are the hormones that characterise the development of the gender specific features. The male hormone testosterone promotes the growth of muscle. The female hormones oestrogen and progesterone promote development of the breasts. As discussed above, women need to maintain healthy levels of body fat to maintain hormonal balance.

Table 1.2 Effects of ageing	
Effects of ageing on body systems	**Associated problems**
Skeletal system	
• Decreased bone density – less calcium in the bones • Calcification (laying down of bone) of the cartilage in joints • Decreased availability of synovial fluid in the joints (fluid which lubricates the joints) • Decreased motor neurones (nerves transmitting messages to the muscles)	• Brittle bones – osteoporosis • Postural problems such as spinal curvatures • Increased likelihood of joint associated disease – arthritis etc • Less effective shock absorption by joints • Stiffer and less mobile joints
Neuro-muscular system	
• Decreased fast twitch muscle fibres (the fibres used during strength training and power activities) • Decreased concentration of myosin and actin (smallest muscles fibres) • Reduced capillarisation (poorer blood supply to muscles) • Increased connective tissue in the muscles • Reduced elasticity in ligaments and tendons • Poor short term memory • Balance impaired • Reduced messages from brain to body due to death of nerve cells	• Reduced movement speed • Less potential muscular strength • Loss of muscle tissue • Less potential muscular endurance • Less flexibility • Stiffer and less mobile joints • Weakened pelvic floor muscles • Forget movement patterns more rapidly • Difficulty stabilising a position and maintaining balance • Reduced body awareness • Reduced movement speed • Increased likelihood of disease of the nervous system (i.e. Parkinson's disease)
Cardiovascular and respiratory systems	
• Decrease gaseous exchange, elasticity of the lungs and flexibility of the thorax • Lower cardiac output and less efficient circulatory system • Reduced capillary network and oxygen delivered to cells • Increased blood pressure	• Reduced oxygen uptake • Lower maximal heart rate and slower recovery rate • Decreased tolerance to fatigue and waste products lactic acid • ncreased likelihood of disease of the cardio-vascular and respiratory system

Adapted from: Lawrence (2004) *The Complete Guide to Exercise in Water.*

Summary

The principles and variables for progressive fitness training discussed in this chapter include:

- Progressive overload – working harder than usual to challenge the body systems
- Frequency – how often we need to train to bring about training benefits
- Intensity – how hard we need to work to bring about training benefits
- Time – how long we need to train for to bring about training benefits
- Recovery time and rest periods – the rest and recovery we need between training sessions
- Reversibility – loss of fitness gains if training ceases
- Specificity – the body adapts specifically to different types of training
- Individual differences – our body type, age, gender will all impact the results of any training programme we follow.

CARDIOVASCULAR FITNESS

The ACSM (2000: 68) describes cardiovascular/respiratory fitness as: 'the ability to perform dynamic, moderate to high intensity exercise for prolonged periods'.

Cardiovascular fitness, cardio-respiratory fitness, aerobic capacity and VO_2Max (maximum oxygen uptake) refer to the body's capacity to produce energy aerobically (with oxygen). Generally the fitter someone is the higher their aerobic capacity or VO_2Max. This means they can work at a higher intensity and continue to provide oxygen to meet the demand before the onset of blood lactate accumulation (OBLA) – this is discussed further in later paragraphs.

Intensity, in relation to cardiovascular fitness, is usually expressed in terms of heart rate. Heart rate monitoring is commonly used to establish and monitor cardiovascular exercise intensity. It is based on the assumption that heart rate is directly linked to the oxygen demands of the muscles (VO_2). An increase in exercise intensity will correspond to an equivalent increase in heart rate and greater demand for oxygen.

The aim of cardiovascular fitness training is to improve the functioning of the heart, lungs and circulatory system to make the whole system more efficient to take in, transport and utilise oxygen and remove waste products (carbon dioxide and lactic acid) at an equivalent rate to meet the demands of the specific activity. When use of oxygen and removal of waste products are equal, energy production is considered to be aerobic and activities can be sustained longer. When the intensity of exercise is such that the build up of waste products exceeds the rate at which oxygen can be supplied, energy production will be anaerobic. These energy systems will be discussed later.

Cardiovascular fitness can be achieved by performing regular activities that elevate the heart rate to an appropriate target level of intensity, which can be sustained for an appropriate duration.

Cardiovascular fitness is an essential component for health. It is also essential for participation in sporting events and for some of the hardcore training programmes described in this book. A stronger heart enables more blood to be pumped around the body in each heart beat (stroke volume). This in turn enables more oxygen to be transported to the working muscles and with less stress being placed on the heart muscle. At the muscular level, an expanded network of capillaries and increased number of mitochondria (cells where aerobic energy is produced) enables more oxygen to be used by the muscles for energy production. In turn, waste products can also be removed more efficiently. Increased strength of the respiratory muscles (intercostals) enables more effective breathing (taking in of oxygen and removal of carbon dioxide). The long term effects of cardiovascular training include a lower heart rate being achieved to pump the same volume of blood to the muscles and a greater uptake of oxygen and more efficient removal of waste products.

Consequently, higher volumes of training (intensity and duration) can be performed by individuals with a trained cardiovascular system. They are able to sustain a comparatively higher

Table 2.1	Training guidelines for cardiovascular fitness
Frequency	3–5 times a week (decrease frequency when working at higher intensity)
Intensity	Between 55–65% and 90% maximal heart rate with different target heart rate zones being used to indicate the working level for different training goals and individual fitness. Equivalent to RPE 12–16 (somewhat hard to hard)
	50–60% of MHR for improving health of untrained individuals 60–70% of MHR for assisting weight management. NB: lower intensity can be sustained for longer and increased duration is often a recommendation for weight management programmes. 70–80% of MHR for developing a base level of cardiovascular fitness from which future training goals can progress. 80–90% of MHR for Improving cardiovascular fitness of highly trained individuals to peak perform. 90–100% for anaerobic performance zone
Time	Between 20 and 60 minutes of continuous or intermittent activity (10 minute bouts accumulated throughout the day), with a minimum of 20 minutes to develop aerobic capacity. ACSM: 2005
Type	Running, swimming, cycling, rowing Any exercise that uses large muscle groups of the body and which can be performed continuously for the sustained duration.
Circuit exercises	Stepping, squatting, skipping, running, jumping etc

working heart rate for longer and effectively manage removal of waste products. Thus, the activities for fitter individuals need to provide sufficient challenge for them to maintain their fitness.

Monitoring intensity for cardiovascular training

Heart rate monitoring and rating of perceived exertion have been shown to have a linear correlation to exercise intensity. That is, as exercise intensity increases, heart rate and rating of perceived exertion will also increase. These methods are therefore used for monitoring intensity.

Heart rate monitoring

To calculate a target heart rate range from which intensity can be monitored it is first necessary to find the maximal heart rate (the most the heart could beat within one minute). An age-predicted formula is used to establish an estimate of the maximal heart rate. However, this needs to be used with some caution as there is a standard deviation (SD) of plus or minus 10–12 beats. This could result in an individual having an actual max HR of 20 bpm higher or lower than the age predicted heart rate (BACR: 2000 in Lawrence & Barnett: 2006).

Calculation for age predicted maximum heart rate for 40 year old client

Age predicted Max HR = 220 – Age

Example: 220 – 40 years = 180 MHR

Calculating target heart rate zone

There are two main methods used for calculating a target heart range which can subsequently be monitored during training.

Percentage of heart rate max (HR max)

This method is calculated by using a specific percentage of HR max. It is calculated using the actual or age-predicted maximum heart rate.

For example if the target heart rate (THR) range is 60–75%:

Step 1: Calculate age-predicted HR max

220 – age = age predicted HR max

Step 2: Calculate target heart rate range

HR max x 0.55 = 55% maximum heart rate
HR max x 0.6 = 60% maximum heart rate
HR max x 0.75 = 75% maximum heart rate

Karvonen or heart rate reserve (HRR)

This formula calculates the heart rate reserve (HRR) to determine a training heart rate. The heart rate reserve is the difference between

---

Example: Client X aged 55

Step 1: Calculate age predicted max HR

220 – 55 = 165

Step 2: Calculate target ranges of 60 and 75% of max HR

Max HR 165 x 0.6 = 99 (60% THR – target heart rate zone)

Max HR 165 x 0.75 = 123.75 (75% THR – target heart rate zone)

Training heart rate zone 60 – 75% of max HR = 99–124

Adapted from: Lawrence & Barnett (2006)

---

the maximal heart rate and the resting heart rate. Resting heart rate is that which would be found after a long sleep or rest. A HRR of 50–70% corresponds to 50–70% of VO_2 reserve. VO_2 reserve is the difference between VO_2 max and resting VO_2. Ideally, a true maximum HR (as obtained from a maximal test) should be used when applying the Karvonen formula. However, a more practical, albeit less accurate approach is to use age predicted max HR.

Step 1: Calculate HRR

Maximum heart rate (or age-predicted heart rate) – resting heart rate = heart rate reserve.

Step 2: Calculate training heart rate (e.g. 50–70% of HRR)

(HRR x 0.5) + RHR = 50% HRR
(HRR x 0.7) + RHR = 70% HRR

Example: Client X Age 55

Step 1:
165 (age-predicted max HR) – 60 (resting heart rate) = 105 HRR

Step 2:
105 HRR x 0.5 = 52.5 + 60 (RHR) = 112.50 = 50% HRR

105 x 0.7 = 73.5 + 60 (RHR) = 133.50 = 70% HRR

Training heart rate zone 50–77% HR max = 112–133

Adapted from: Lawrence & Barnett (2006)

Monitoring heart rate response to exercise

Manual heart rate monitoring requires a reasonable level of skill and competence. It is therefore comparatively less accurate than using heart rate monitoring devices. It may be more appropriate to use heart rate monitoring in conjunction with other methods, such as RPE.

Rating of perceived exertion (RPE)

Borg (1998) in ACSM (2007: 77) developed the RPE 6–20 and CR–10 to enable subjective quantifying of exercise intensity. The 6–20 scale is designed for rating overall feelings of exertion and is generally used for steady state aerobic activity. The CR–10 scale is designed for rating more individualised responses such as breathlessness and pain.

When using RPE, the client is encouraged to focus on the sensations of physical exertion (breathlessness, strain and fatigue in muscles) and then rate their overall feelings of exertion against the scales. The more experienced a client becomes at detecting and rating sensations, the

Fig 2.1 The Borg Rating of Perceived Exertion (RPE) 6–20 scale and the Borg Category Ratio–10 scale

RPE Scale		CR10 Scale		
		0	Nothing at all	'No P'
6	No exertion at all	0.3		
7	Extremely light	0.5	Extremely weak	Just noticeable
8				
9	Very light	1	Very weak	Light
10		1.5		
		2	Weak	
		2.5		
11	Light	3	Moderate	Strong
12		4		
13	Somewhat hard	5	Strong	
14		6		
15	Hard (heavy)	7	Very strong	
16		8		
17	Very hard	9	**Extremely strong**	**'Max P'**
18		10		
19	Extremely hard	11		
20	Maximal exertion			
		•	Absolute maximum	Highest possible

Borg G (1998) in ACSM 2006: 77

more closely the ratings correlate with the exercise intensity.

The ACSM (2000: 78) indicate that a cardiovascular training effect and the threshold at which blood lactate starts to accumulate is achieved at a rating of between 'somewhat hard' to 'hard', which equates to a rating of 12–16 (RPE) or 4–6 (CR –10) on the respective scales. Fatigue levels (where lactate builds and cannot be removed at a sufficient rate, requiring intensity to be reduced to maintain the activity) equate to ratings of between 18–20 and 9–10 on the respective scales.

A key consideration would be that the actual intensity of exercise to achieve these ratings will be different between subjects, depending on fitness. For example, an untrained person may subjectively rate intensity at 12–16 when walking at a moderate pace on a very slight incline. Whereas, a trained individual may rate intensity at 12–16 while running at a fast pace on a higher incline. The point being, the type of activity to achieve the ratings will be determined by individual factors, which include fitness and skill level. What an untrained client finds subjectively hard, a trained client will find subjectively easy.

Aerobic and anaerobic (the energy systems)

The energy currency used by the body to produce energy comes in the form of a chemical called adenosine tri-phosphate (ATP). ATP derives from the break down (via digestion) of nutrients (carbohydrates, fats and protein) we obtain from our diet. ATP is stored in all cells of the body that require energy to function – this includes muscle cells.

When the demand for energy is made, one of the phosphates break away to produce energy. This leaves another chemical – adenosine di-phosphate (ADP) – which needs to be converted back to adenosine tri-phosphate before further energy can be produced.

$$ATP - P = ADP$$

ATP is stored in limited amounts in the muscle cells and needs to be continually broken down and re-synthesised inside the muscle via the energy systems. During times of rest, this breakdown and re-synthesis occurs at a slower rate. During exercise and activity, demands for ATP are greater and this requires energy to be produced at a faster rate.

The three energy systems that enable the re-synthesis of ATP are:

- The creatine phosphate system
- The glycogen or lactic acid system
- The oxygen system

The first two systems operate without the presence of oxygen and are classified as anaerobic. The third energy system requires the presence of oxygen and is therefore classified as aerobic.

Creatine phosphate system

Creatine phosphate (CP) is another chemical produced by the breakdown of nutrients from the diet. It is stored in the muscles and provides the immediate source for re-synthesising ATP. The phosphate that is attached to the creatine releases and joins with ADP to remake ATP, which in turn, releases the phosphate to produce energy.

$$CP + ADP = ATP + C$$

Stores of creatine phosphate are limited within the muscle. Thus, energy can only be supplied for a short duration, usually no longer than 6–10 seconds. This system would be used as the predominant source during immediate bursts of high to maximal intensity that last for a short duration. Example activities would include athletic track and field events that involve sprinting (100m), throwing (javelin) and jumping (long jump).

Within a circuit training session, explosive power movements such as tuck jumps and burpees would most likely challenge this system. Rest intervals between higher intensity exercises will need to be longer to enable replenishment of ATP-CP stores and reduce the accumulation of lactic acid.

Glycogen/lactic acid system

Glycogen is produced by the break down of the nutrient carbohydrate. It is stored in the muscle and liver. Stored glycogen can be broken down to reform ATP for bursts of higher intensity activity. However, without the presence of oxygen in the breakdown of glycogen, the waste product called lactic acid is produced. When lactic acid accumulates it contributes to fatigue and is experienced as a burning sensation in the muscles. Exercise intensity will need to decrease or be stopped to assist removal of lactic acid via the circulatory system.

Glycogen + ADP = ATP + Lactic Acid

The glycogen system is used predominantly during high intensity activities that usually last for no longer than 90 seconds. McArdle, Katch & Katch (1991: 125) suggest that: 'The most rapidly accumulated and highest lactic acid levels are reached during exercise that can be sustained for 60–180 seconds. Athletic track events that use this energy system would be the 400 and 800 metres. It would also be used, when pushing to finish the final stage of a mile run'.

According to McArdle et al (1991: 125): 'Lactic acid begins to accumulate and rise in an exponential fashion at about 55% of the healthy, untrained subject's maximal capacity (VO_2 Max) for aerobic metabolism'. As exercise intensity increases, so too does the level of lactic acid. For trained individuals, the accumulation of lactic acid generally occurs at a higher level of their maximal capacity (VO_2 max). Trained endurance athletes frequently perform at intensities equivalent to between 80% and 90% of their maximal capacity (VO_2 max). This indicates that the level at which the onset of blood lactate (OBLA) occurs can be improved through training (see page 18 for physiological adaptations to aerobic training that may contribute to OBLA occurring at higher exercise intensities).

The point at which lactic acid accumulates, the blood lactate threshold, is commonly referred to as the anaerobic threshold.

Oxygen system

This energy system uses a combination of nutrients as fuel. These nutrients are broken down in the presence of oxygen (aerobic) to produce energy.

- Carbohydrates: obtained from pasta, rice, potatoes (stored as glycogen).
- Fat: obtained from dairy products – butter, cheese, milk (stored as adipose tissue).
- Protein: obtained from meat and vegetable sources. These are used primarily for growth and repair and only used for energy when other fuels are depleted.

Glycogen + Fats + protein + oxygen + ADP = ATP + Carbon dioxide + water

Aerobic energy production occurs in specialised cells within the muscle called mitochondria. These cells contain the enzymes needed to use oxygen. A long-term effect of regular aerobic cardiovascular training is that a greater number of mitochondria develop within the muscle and they also increase in size. A further effect is that the capillary beds in the muscle expand. This allows more oxygen to be delivered to the working muscles and used for energy production. In addition, these adaptations provide for more effective removal of waste products that accumulate (lactic acid). These muscular adaptations from cardiovascular training are cited as being potentially responsible for the onset of blood lactate accumulation (OBLA) occurring at higher levels of intensity among trained individuals. 'The exercise intensity at the point of OBLA is a consistent and powerful predictor of performance in aerobic exercise.' (McArdle, Katch & Katch: 1991: 281). The benefits of training the aerobic energy system will therefore potentially contribute physiological improvements (increased size and number of mitochondria and increased capillaries at local muscular level) that enable greater levels of work. A further benefit is that the body becomes more efficient in using its fat stores. Aerobic cardiovascular training is often a key prescriptive feature of weight management and fat loss training programmes.

A possible limitation of this system is that it takes slightly longer to engage. The heart rate and breathing rate will need to increase and the capillary network will need to dilate to enable the delivery of oxygen to the working muscles. These processes take a few minutes, hence during exercise the warm-up process is essential to enable time for these responses and physiological adjustments to occur. One major advantage of this system is that greater amounts of ATP can be produced and these increased levels can be maintained for longer periods when working at a steady pace. In addition, the waste products produced – carbon dioxide and heat/water – are comparatively easy for the body to remove through exhalation and sweating.

Activities that use the aerobic system are those that can be maintained longer but at a lower to moderate intensity. Athletic events such as the marathon and other long distance endurance running, swimming and cycling events mainly use this system.

Energy systems in action

During most circuit training sessions, the energy systems interweave. At different times throughout the workout, different energy systems may be more predominant depending on the intensity and duration of the activity and the fitness level of the individual performing it.

At the start of the session, all systems will be engaged, with the anaerobic system being most active, until the circulatory system has time to respond (capillary dilation etc). If the intensity of the warm-up and preparatory phase is appropriate, the aerobic system will become the predominant system. If the intensity of the warm-up is too high, exercise intensity will need to be reduced to maximise the aerobic system.

During the main circuit, the intensity of specific exercise stations and the fitness level of the individual will determine the energy system used. High intensity stations involving sprinting, jumping, heavy lifting will use the anaerobic system more predominately. Lower intensity stations and recovery periods will allow the aerobic system to be used more predominately.

At the end of the session, when the body is cooling down and recovering, the aerobic energy system will be most active.

Table 2.2	Energy system summary		
Energy system	Creatine phosphate	Lactic acid/glycogen	Oxygen
Fuel used	Creatine phosphate	Glycogen	Glycogen, fat protein
ATP production	Very limited	Limited	Unlimited
Engagement	Very rapid – immediate	Rapid	Slow (cardiovascular system need to engage to circulate blood/oxygen).
Muscle fibre	Fast twitch	Fast oxidative glycolytic (FOG)	Slow twitch
Fuel stored	Muscles	Muscles and liver	Glycogen stored in muscle. Fat stored as adipose tissue.
Amounts of fuel available	Limited. Only a few seconds.	Moderate supply – from a few seconds to up to 2–3 minutes.	Glyocgen moderate Fat more plentiful System can sustain activity for as long as fuel available.
Intensity, duration and type of activity	High intensity Short duration 100m sprint. Throwing and jumping events. Strength training	Moderate to high intensity 400m sprint Anaerobic endurance – 8–25 repetitions approximately	Low to moderate intensity. Longer duration Marathon running Long distance swimming and cycling. Circuit weight training
Waste products	Creatine	Lactic acid accumulation inhibits muscle contraction.	Carbon dioxide exhaled Heat generated, body produces water/sweat to regulate temperature.
Limitations	Short supply of creatine phosphate	Lactic acid build up	Takes longer to engage

Adapted from: Lawrence & Hope (2004)

Table 2.3	Energy systems used in different sporting activities	
Sporting events	Aerobic – with oxygen	Anaerobic – without oxygen
Marathon run	100%	0%
Cross country distance run 10 km Run/race	90%	10%
3000 metre run/race	80%	20%
400 metre swim/race Rowing 2000 metres	60–70%	30–40%
1500 metre run Hockey game Football game Rugby game NB: position of player and activity levels will alter energy systems used.	50%	50%
800 metre run	40%	60%
Boxing	30%	70%
100 metre swim	20%	80%
200 metre sprint	10%	90%
100 metre sprint or shorter sprints	0%	100%

Adapted from: Davis, Roscoe, Roscoe, Bull (2005) & Davis, Kimmet, Auty (1986)

Variables for progressive cardiovascular training

There are three main phases for developing cardiovascular conditioning programmes (ACSM: 2000: 154). These include:

Initial conditioning stage

Four weeks – building a light endurance base while establishing appropriate frequency, intensity and duration with the aim of promoting adherence for non-exercisers and minimise the risk of muscle soreness, injury and discomfort.

Improvement stage

Four to five months – building the volume of exercise by increasing intensity, duration and speed.

Maintenance stage

Beyond five months – maintaining fitness levels and working towards specific goals.

These guidelines should be applied to reflect the specific needs of the individual, their existing level of fitness and training goals.

The intensity of cardiovascular training is monitored using heart rate and RPE. The target heart rates for specific fitness goals are listed in Table 2.4.

There are a number variables that can be modified to alter the intensity (heart rate and RPE) and thus be used to progress cardiovascular training. These include:

• Rate/speed
• Resistance/weight and range of motion
• Repetitions/rest

Table 2.4	Target heart rate zones
% Max HR	Target heart rate zone
100%–90%	Anaerobic performance zone
90%–80%	Improving cardiovascular fitness of highly trained individuals to peak performance levels. Onset of blood lactate (OBLA) zone.
80%–70%	Developing a base level of cardiovascular fitness from which future training goals can progress.
70%–60%	Assisting weight management NB: lower intensity can be sustained for longer and increased duration is often a recommendation for weight management programmes.
60%–50%	Improving health of untrained individuals.

Adapted from: Davis, Roscoe, Roscoe, Bull (2005) & Davis, Kimmet, Auty (1986)

Rate/speed

All cardiovascular exercises need to be performed at a sufficient speed to provide training benefits. Working at a slower pace will generally be easier than working at a higher pace. However, if exercises are too slow they may lose their rhythmical nature. If they are too fast, control may be lost. The speed or rate must be comfortable and sustainable for an appropriate duration. If the training focus is to play with speed (fartlek training) throughout the workout, then a variety of paces can be used and will achieve specific benefits. Fartlek training is discussed later in this chapter.

Cardiovascular machine training offers the easiest way of changing and monitoring speed/pace (see Table 2.5).

Table 2.5	Cardiovascular machines – speed variables
Cardio vascular machines	Variant: rate/speed measure
Rowing machine	Strokes per minute (SPM).
Cycling machine	Revolutions per minute (RPM).
Treadmill	Miles per hour (MPH) or kilometres per hour (KPH).
Step machine	Number of steps or floors climbed.
Cross trainer	Number of stride revolutions.

NB: some new machine models may have KPH to replace stride revolutions.

From: Lawrence & Barnett (2006)

Resistance/weight and range of motion

Body weight provides resistance. More effort is required to move a larger, heavier body frame and move against the resistance of gravity. Most cardiovascular machines offer a variable resistance workload. The intensity or workload can be altered by changing the level of resistance of the machine.

During circuit training programmes body weight can be used to manipulate the intensity by moving the centre of gravity through a larger range of motion (see Table 2.6).

Repetitions/rest

Examples of manipulating the repetitions and rest periods include:
- Working for longer to perform a specific exercise within a circuit – for example shuttle runs
- Increasing the time on a circuit station
- Reducing the rest time or making the rest more active, for example squats as main circuit station with active rest of jogging before moving to next station

Interval and fartlek methods (discussed later) can be used to prescribe work and rest times (work: rest ratio) within the exercise/activity.

As fitness develops the following can be manipulated:
- Increase work times on stations
- Increase the number of work intervals (more stations)
- Work for longer
- Work more frequently (more work intervals)
- Reduce rest intervals

Training methods

Continuous training

Continuous training involves maintaining the heart rate at the appropriate level of intensity (target heart rate zone), below the anaerobic threshold for at least 20 minutes. The aim of continuous training is to develop aerobic capacity. This training is used during walking, jogging, running, swimming, cycling and

Table 2.6	Range of motion
Method	Example exercises
Bending, deeper	Low impact squatting, travelling and body weight movements can be made harder by deepening the knee bend. Gym mats can be used for running and jumping activities. They provide less stability and the muscles have to work harder to move from the unstable surface and also maintain balance.
Impact	Increasing the height of jumping movements Jumping up on to steps Including explosive exercises – tuck jumps
Incline	Increasing the number of risers on a step block
Travelling	Increasing the distance travelled – shuttle stations being placed at further distance apart.

Adapted from: Lawrence and Barnett (2006)

rowing where it is relatively easy to monitor and maintain a continuous level of intensity.

Interval training

Interval training is the performance of shorter, higher intensity work periods between lower intensity rest and recovery periods.

Interval training enables all energy systems to be trained. Although, the predominant system will be determined by the duration, intensity and type of exercise included in the work interval as well as the duration, intensity and type of activity included in the rest interval and the fitness level of the individual.

By its nature, circuit training tends to use an interval approach. There are a number of specific work/circuit stations, with planned active rest phases between each station.

The work rest ratio can be: 1:2 or 1:3 (or other variations, determined by the energy system and recovery time needed. As a general rule, the higher the intensity, the longer the rest time between intervals).

For example:
Work station 1: 30 seconds squat thrusts
Rest: 60–90 second steady jog
Work station 2: 30 seconds press-ups
Rest: 60–90 seconds steady jog
Exercises designed to challenge the ATP-CP system will need a shorter work time and a longer and less active rest phase to enable replenishment of ATP and CP. An appropriately calculated rest interval slows down the build up of lactic acid, enabling the overall workload to be maintained.

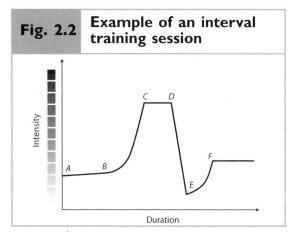

Fig. 2.2 Example of an interval training session

Fartlek training

Fartlek training involves varying the speed during a training session. A continuous training session (walking/jogging) is established and short bursts of speed are integrated every couple of minutes (running or sprinting).

Example 1 (see Fig. 2.2)
A circuit session using a command approach. A continuous pace is established (A, B and F). Every two to three minutes a command is shouted to perform a different exercise that manipulates the speed (C and D). The intensity is then reduced to allow recovery (E).

Example 2
Position the work stations at two ends of a sports hall.
Establish a continuous pace (walking or jogging) between stations.
Maintain the continuous pace for a few minutes.
Perform a work station.
Continue as above, varying work stations.

Example 3
Position the work stations at two ends of a sports hall
Sit-ups
Walk one length of hall
Sit-ups
Jog two lengths of hall
Press-ups
Jog three lengths of hall
Continue with a range of different exercise stations and different levels of running and walking (race walk, sprint, steady jog etc to challenge different energy systems).

Summary

The key learning points related to cardiovascular training discussed in this chapter include:

- CV training improves the fitness of the heart, lungs and circulatory system
- Running, stepping, swimming and cycling are example of CV exercises
- CV training needs to be performed on 3–5 days a week, for 20–60 minutes (a warm-up should precede the workout and a cool down should end it)
- CV exercises need to be performed at an appropriate intensity within the target heart rate zone of 55–90% of MHR
- Intensity can be monitored by using observation, RPE, heart rate monitoring or talk test
- The three energy systems (creatine phosphate, glycogen and oxygen) interweave throughout a session depending on the intensity and duration of the activity and fitness level of the group
- Different approaches to CV training include: continuous, interval and fartlek.

FLEXIBILITY TRAINING

3

Flexibility is the ability of our joints and muscles to move through their full potential range of movement. Flexibility is important for maintaining correct posture and alignment and carrying out daily activities. It is also essential for performance in specific sports (gymnastics and dance). Without sufficient levels of flexibility, there is the potential for tissue damage when the joint structures (muscles, ligaments and tendons) move beyond 'a joint's shortened range of motion' (ACSM. 2000: 85).

Factors affecting flexibility

The range of motion available depends on a number of factors. These include temperature, joint structure and surrounding tissues (tendons, ligaments), muscle connective tissue, age, gender etc.

Temperature

Warm muscles are more pliable and will stretch more effectively. It is therefore essential that some active warming occurs before stretching muscles. Time of day may also influence flexibility potential. Generally, people report feeling more flexible in the evening than in the morning. This may be due to daily activities that have provided some movement to the body throughout the day.

Joint structure

Different synovial joints have different shapes and consequently allow different movements and ranges of motion to occur, in accordance with their shape and structure.

Ligaments, tendons, hyaline cartilage and connective tissue

Joints are surrounded by specific structures that affect range of motion. These include:

- ligaments – which attach bone to bone and maintain joint stability
- tendons – which attach muscle to bone
- hyaline cartilage – which forms the ends of the bone and allows smoother movement at joints
- connective tissue – which is the material between the cells of the body that gives tissues, including muscles, form and strength.

Ligaments

The function of ligaments is to hold the bones together and provide stability for the joint. Ligaments are made of dense regular collagen fibres bundled tightly together which gives them strength to resist excessive movement. Movements that place the joint out of its natural range of motion may cause injury to the ligaments. This reduces the stability of the joint. Any severe damage to ligaments would require surgical intervention. Ligaments do not have a rich blood supply and consequently do not heal naturally.

Table 3.1	Recommended training guidelines: flexibility
Frequency	2–3 times per week minimum 5–7 times per week ideal
Intensity	Point of mild tension/tightness in the belly of the muscle at the end of the range – not pain. 2–4 repetitions of each stretch can be performed
Time	Static stretching – 15–30 seconds hold PNF stretching – 6 second contract followed by 10–30 second assisted stretch
Type	Static or PNF techniques for all major muscle groups Slow and controlled performance progressing to greater ranges of movement.

NB: to stretch all muscles using PNF techniques would demand a longer time for stretching. Muscles would also need to maintain an appropriate (warm) core temperature to stretch effectively. With these considerations in mind, it may be necessary to prioritise PNF stretches within a particular training session.

It is also essential to give consideration to the specific flexibility needs of the individuals (current flexibility levels and flexibility needs for their specific sporting activity).

Adapted from: ACSM (2000: 158) and ACSM (2006: 162)

Tendons

Tendons attach muscle to bones across joints. They are made from all the muscle fibres and connective tissue around them being compacted together. Tendons can be felt at various points around the body. One example is the Achilles tendon at the back of the heel. Tendons feel like tight metal rods. They provide around 10% of the total resistance to movement around a joint.

Hyaline cartilage

Hyaline cartilage forms the ends of the bones that meet to form synovial joints. Cartilage prevents the bones from rubbing against each other during movement. Hyaline cartilage is susceptible to calcification (laying down of calcium) and can ossify into bone. This has implications for joint movement and can result in arthritis. (Lawrence & Barnett, 2006).

Connective tissue

Muscles are made up of bundles of muscle fibres. At each layer of the muscle, there is connective tissue, which surrounds the tissues (epimysium, endomysium, perimysium). Connective tissue is made from collagen, which is relatively inelastic. This connective tissue accounts for around 40% of the total resistance to movement around the joint.

Static stretching when very warm can influence the length of connective tissue. As muscle temperature increases the stiffness of the connective tissue decreases and extensibility increases.

Gender

Although cited as a factor that may affect flexibility, there is no conclusive evidence to substantiate the generalisation that women are more flexible than men. There are, however, some anatomical differences between genders that may influence flexibility. Women have broader and shallower hips and thus they potentially have a greater range of movement than men in this area.

A further consideration may be the activities we participate in during formative years, which may contribute to maintaining a range of movement. Sedentary lifestyles contribute to a reduced range of movement and more active lifestyles (depending on specific activities) will contribute to maintaining the range of motion. Some sporting activities such ballet and gymnastics offer a greater focus on range of motion than other sports, such as football and running, where other components of fitness may be emphasised. If insufficient attention is paid to stretching, the muscles and joints will lose their range of motion, contributing to a change in the skeletal alignment and posture. This in itself offers risk of injury through poor posture and the possibility of low back pain.

Age

Flexibility can be developed at any age but the rate of improvement will not be the same. It is generally thought that the greatest improvement in flexibility occurs between the ages of 5–7. However, caution must be taken during the growth spurt (girls of approximately 10–12 years old and boys of approximately 12–14 years) because there is a tightening of muscles at these ages. The muscle is already being stretched due to the increase in bone length and joint size. The soft tissue does not keep pace with this growth rate and this increases the risk of injury.

For reference, there are no specific flexibility guidelines for working with children and young adults. The standard adult guidelines are therefore used.

Physiology related to flexibility and stretching

A muscle is considered to be stretching (lengthening) when the points of attachment, the origin and insertion, move further apart and the muscle relaxes.

When a muscle stretches sensory receptors (proprioceptors) within the muscles and tendons monitor the changes occurring to the body position and communicate this information back to the central nervous system (CNS). These proprioceptors form part of the peripheral nervous system (see chapter 5). They include:

- The muscle spindles (stretch reflex)
- Golgi tendon organs.

When activated, these receptors deliver the information they gather back to the CNS, spinal cord and brain via the afferent nerves. The brain responds to the information received and sends a message back to the muscles (via the efferent nerves) instructing the muscles to either relax or contract a little more or a little less so that correct body alignment is achieved.

The stretch reflex is stimulated when there is a change in muscle length. The muscle spindles are fibres located deep inside the muscles and surrounded by nerves. They register both the speed and amount of lengthening or stretch that is occurring. The faster the speed of stretch the faster the speed of messages between the muscles and CNS.

When a muscle lengthens too quickly (ballistic stretching), the stretch reflex is activated (via the CNS). The CNS informs the muscle being lengthened to contract to prevent injury to it and the surrounding tissues. Motor neurons (efferent nerves) going to the opposite (antagonist) muscle are also inhibited so that it relaxes and cannot interfere with the agonist muscle shortening. The stretch reflex is, therefore, rather like a protective mechanism that prevents the risk of injury to the muscle.

For example:
The stretch reflex can be observed in action when someone falls asleep while sitting upright. As the head falls forward into a flexed/bent position, the muscles at the back of the neck lengthen quickly and are overstretched. The stretch reflex causes the head to jerk back up into extension, usually waking the person.

During ballistic and bouncy stretching, the stretch reflex is continually activated. This is one factor that increases the potential for injury during ballistic stretching. The warnings for over stretching are ignored and the repetitive bouncing and momentum can potentially take the muscle/joint beyond its existing range of motion, without adequate control. This may cause over stretching of the tissues that stabilise and surround the joint (ligaments, tendons and muscles). Ballistic and other types of stretching are discussed later.

The Golgi tendon organs (GTOs) are located in the tendon. They monitor and relay information about the level of tension in the muscle. Tension can be created by excessive stretching or contraction of the muscle. The GTOs respond to fast and forceful contraction. They function in the opposite way to the stretch reflex in that they inform the muscle to relax.

For example:
If a weight lifter is bench pressing and lifting a weight that is too heavy, excess tension may build up causing the muscle to eventually give in to the resistance and relax. The consequences of this during weight lifting can be dangerous – hence the recommendation of a spotter.

The GTO response is exploited during PNF stretching. PNF and other types of stretching are discussed later.

Training methods for flexibility and stretching

The ACSM (2000) suggest that stretching can be incorporated into both the warm-up and cool down of an exercise session. They recommend that an active warm-up precedes any warm-up stretching.

Dynamic stretching

Dynamic stretching is where there is movement through the full range of motion. The muscle and joint are lengthened to an extended position and then moved out of the stretch. A number of repetitions can be performed. Movements need to be controlled to prevent any ballistic action. A range of dynamic stretches are illustrated and explained in part 2, chapter 6.

Static stretching

Static stretching is achieved by lengthening the muscle slowly to a point where a mild tension is felt in the belly of the muscle. The stretch is then held for an extended duration, usually between 10–30 seconds. This method is more commonly recommended because there is a lower risk of injury and muscle soreness.

Active stretching

Active stretching occurs when the antagonist muscle contracts to achieve a stretch (see Fig. 3.1a).

The advantage of this stretching technique is that the existing range of motion will not usually be exceeded. The disadvantage is that the range of motion achieved is reliant on the strength of the antagonist muscle that is contracting to bring about the stretch. A further disadvantage may be the isometric contraction of the antagonist muscle, which limits the comfort of the stretch position.

Passive stretching

Passive stretching occurs when both the agonist and antagonist relax. This is achieved by supporting the stretch position.

> For example:
> In the example listed below for the quadriceps. Using the hand to hold the foot will enable the hamstring (antagonist) muscle to relax and will potentially bring about a greater range of motion (see Fig. 3.1b).

The key consideration during passive stretching is to be aware of sensations in the body and be alert to activation of the stretch reflex (experienced as the muscle is shaking or feeling tight). If the stretch reflex does not ease, the range of motion should be reduced, to ensure safe stretching.

Ballistic stretching

Ballistic stretching involves lengthening the muscle by using repetitive bouncing movements. There is a risk of injury and of delayed onset muscle soreness (DOMS) associated with this type of stretching. This risk increases when the bouncing movements are faster, and greater

Fig. 3.1a Active stretch of quadriceps muscle (hamstring contracts to lift foot)

momentum forces are applied. For these reasons, we do not recommend ballistic stretching.

Proprioceptive neuromuscular facilitation (PNF)

PNF stretching involves taking the muscle to the end of the range of movement and then contracting the muscle isometrically (static) against a resistance that does not move. There is an alternation between contraction and relaxation of the prime mover and antagonist muscles. PNF can produce great improvements in flexibility.

Fig. 3.1b	Passive stretch of quadriceps muscle (hand holds ankle)

However, there is a risk of muscle soreness. This technique requires more time than other methods and the partner used must be competent in the practice of PNF techniques. A range of partner stretches is listed in part 2, chapter 7.

PNF stretching exploits the response of the Golgi tendon organ. It involves stretching the muscle and joint slowly to the end of the range of movement and then providing a strong static/isometric contraction against a resistance/external force usually provided by a partner for six seconds. The increased tension causes the inverse stretch reflex to be activated. This results in relaxation of the muscle and

enables the joint to move further. Holding this extended position allows the connective tissue to lengthen (collagen creep) and although it will shorten back when the stretch is released, some increased length will remain.

Contract relax PNF method

This involves contracting a specific muscle (usually with partner assistance), relaxing that muscle and then moving further into the stretch when the muscle has relaxed.

For example:
Lying hamstring stretch

1 Raise the leg towards the chest to achieve a stretch of the hamstrings
2 Contract the hamstring muscle by pushing against a partner for six seconds
3 Relax the muscle
4 Ease further into the stretch with partner assistance

Contract relax antagonist contract PNF method (CRAC)

This involves contracting and relaxing the muscle as above with the addition of a secondary contraction of the antagonist muscle (following the relax phase) to bring about a further increase in range of motion.

Self-PNF stretching

An alternative to partner PNF stretching is to perform it by yourself. Instead of using a partner, an isometric contraction can be achieved by using any of the following: a towel, the wall, a yoga strap or just the hands/arms to contract against.

This method may be more appropriate until a good rapport and level of trust has been established between exerciser and trainer. Self-PNF

Fig. 3.2 **Hip flexor stretch: range of motion difference between standing and kneeling**
(a) Kneeling
(b) Standing

techniques can also be used when working with groups.

The process and timings are the same as with partner PNF stretching already described. Muscles groups such as hamstrings, adductors, calves and pectorals are those that are ideally suited to self-PNF stretching.

Variables for progressive flexibility training

There are specific variables that can be altered to change and adapt the intensity of specific stretches. These are:

- Range of motion
- Balance and stability of stretch positions
- Isolation of muscles being stretched
- Type of stretching (static, PNF etc)

Fig. 3.3 **Lying hamstring stretch: range of motion difference**
(a) hand holding thigh
(b) towel around foot to assist

Range of motion

Flexibility training needs to begin from the range of motion an individual already has. This will differ between joints and between individuals. Static and passive stretches in supported positions will offer greater control for performing the stretch and achieving an appropriate range of motion.

Gravity can be used to assist stretches. For example: lying on a step and letting the arms drop horizontally to the sides, level with the shoulder, will offer a passive stretch of the pectorals and anterior deltoid.

Towels can be used to support levers and trainer-assisted stretches can be used to assist the range of motion and to support the stretch position.

Balance

Some stretch positions require greater levels of balance to achieve an effective stretch.

Using a wall to provide balance, using floor based positions, using step or chair based stretch positions are all appropriate ways to adapt stretch positions and assist performance.

Fig. 3.4 Pectoral stretch lying on step

Fig. 3.5 Lying hamstring stretch with towel

Fig. 3.6 Standing quad stretch using towel

Isolation

Isolating the muscle stretching (where possible) will make the stretch position easier.

For example:
A straddle adductor stretch will require greater flexibility of other muscles, including hamstrings, than a feet together adductor stretch. Both seated and straddle positions demand strength of the core muscles and flexibility around the hips and back to sit in an upright position.

| Fig. 3.7 | Pectoral stretch using door frame | Fig. 3.8 | Standing calf stretch using wall |

| Fig. 3.9 | Hamstring stretch (a) Floor based (b) Seated – decreased range of motion |

Fig. 3.10 **Adductor stretches**
(a) Seated straddle
(b) Seated adductor

Summary

The key points discussed in this chapter include:
- Flexibility training improves the range of movement available at the joints/muscles
- To improve flexibility we need to stretch the muscles regularly (between 2–7 days a week)
- The muscles need to be lengthened to a point where a mild tension is felt in the muscle
- Static stretches are those recommended for general populations
- Flexibility is influenced by age, gender, joint structure, body type etc
- The different methods of stretching include: static, active, passive, dynamic, ballistic and PNF.

MUSCULAR FITNESS (STRENGTH AND ENDURANCE)

4

Muscular fitness describes a balanced combination of muscular strength and muscular endurance. Muscular strength and muscular endurance represent polar ends of a continuum.

Muscular strength is the ability of our muscles to exert a near maximal force to lift a resistance for a short duration (low repetitions). Muscular endurance requires a less maximal force to be exerted, but for the muscle contraction to be maintained for a longer duration (higher repetitions). Muscular fitness represents an optimal level of fitness for functional purposes (between the two polar extremes). Sports people and athletes may need to train for either muscular strength or muscular endurance depending on their fitness goals and the sport they participate in.

Different training regimes will bring about different benefits and physiological adaptations. These are outlined in Table 4.1

Benefits of muscular fitness training (strength and endurance)

Muscular fitness is essential for health. Some of the benefits include:

- Increased bone density and decreased risk of osteoporosis
- Increased lean tissue and improved metabolic rate (calorie burning)

- Increased strength of ligaments and tendons, reducing injury risk
- Improved posture and stability, reducing risk of low back pain
- Improved ability to carry out daily activities (lifting, pushing, pressing, carrying) thus improving physical esteem
- Improved glucose tolerance, reducing risk of type 2 diabetes

Muscular fitness, strength and endurance are also essential components of sport and athletics. A good all round level of muscular fitness is probably required for participation in sport.

Different sports will also require some specific training emphasis (strength or endurance) depending on the event. For example: power lifters would need to focus on power and strength training while endurance athletes would need to focus on endurance. This is because specific training brings specific gains.

Training consideration for specific sports would need to include:

- The muscle(s) being used (upper limb – lower limb – trunk)
- The type of muscle contraction and range of motion (isometric – isotonic – joint angle)
- The speed of contraction (fast – slow)
- The resistance and repetitions (strength – endurance)
- The energy system

The principles relating to muscle work will now be explored further.

Table 4.1	Physiological adaptations to muscular strength, fitness and endurance		
Muscular strength	**Muscular fitness**	**Muscular endurance**
1–6/8 repetition maximum (RM)	8–12 repetition maximum (RM)	12–30 repetition maximum (RM)
Heavy resistance 100%–75% of 1RM	Sub maximal resistance 75%–60% of 1RM	Lighter resistance Below 60% of 1RM
Anaerobic Creatine phosphate (ATP-CP)	Anaerobic Glycogen/lactic acid	Anaerobic–aerobic Glycogen/anaerobic – oxygen
Fast twitch	Fast twitch – FOG	FOG – Slow twitch
Power lifting	Functional	Endurance training
Physiological adaptations Long term hypertrophy (increase in size/bulk) of the fast twitch muscle fibres.		**Physiological adaptations** Muscle toning and some hypertrophy. These fibres do not have the same growth potential as fast twitch fibres.	
Neuromuscular efficiency Initial strength gains are usually a result of more efficient muscle fibre recruitment, which in turn results in a stronger muscle contraction rather than muscle growth.		**Neuromuscular efficiency** Initial adaptations will be in the efficiency of fibre recruitment.	
Tendons and ligaments The strength of ligaments and tendons improves contributing to better joint stability.		**Capillaries and mitochondria** The muscle cells respond to the build up of lactic acid increasing the size and number of capillaries and mitochondria.	
Cautionary note: Maximal strength training is not recommended for groups. This type of training should be taught in a personal training environment with the provision of spotting.			

Principles relating to muscle work

Muscles attach to the skeleton at different points. The different ends of the muscle are called the origin and insertion. The origin of the muscle is usually fixed and doesn't move. It is usually positioned closer to the body centre.

The insertion is the part that moves and is usually positioned furthest from the body centre.

Muscles have to cross over and pull on joints to create movement (see Table 4.3 below for specific joint/muscle movements). They work in pairs. As one muscle contracts (prime mover) the opposite muscle will relax (antagonist).

Table 4.2	Recommended training guidelines: muscular fitness
Frequency	2–3 days a week for the same muscle groups on non consecutive days More frequent training and increased sets and repetitions are appropriate for trained individuals and may provide increased strength gains.
Intensity	Minimum guidelines 1 set of 3–20 reps (3–5, 8–10, 12–15) 8–12 repetitions to point of volitional fatigue Increased repetitions for muscular endurance with lighter resistance (10–15 repetitions) Increased resistance for muscular strength with lower repetitions (maximal or sub maximal resistance recommended 1–6/8 RM). 8–10 exercises covering all muscle groups exercising whole body (back, arms, legs, abdominals, chest, hips, shoulders). Full range of motion – isotonic Moderate to slow speed to enable concentric and eccentric control. Caution advised when focusing on eccentric contraction due to increased muscle soreness.
Time	1 hour maximum duration to promote adherence
Type	Correct exercise technique Correct breathing to minimise potential for increases in blood pressure Free weights Fixed resistance Body weight Exercise bands Other resistance equipment described in part 4, pages 106–176.

Adapted from: ACSM (2000) and ACSM (2006).

Other muscles will also be working in either a fixating or synergistic role.

Fixator: these are muscles that contract to fix/hold the origin of the prime mover in place.

Example: Fixator
During a bicep curl muscle around the shoulder joint and shoulder girdle would be contracting to fixate.

Synergyst: these are muscles that contract to prevent unnecessary movement at other joints.

Example: Synergist
During a bicep curl the muscles of the forearm would contract to prevent movement at the wrist joint.

To move the body frame (skeleton), the muscles have to contract to overcome the force of gravity, body weight and any external weight applied (free weights, fixed weights etc). Muscles contract in the following ways:

Concentric: where the two ends of the muscle (origin and insertion) move closer together to

Table 4.3	Joint/muscle actions			
Muscle	Position	Joints crossed	Prime action	Example exercise:
Gastrocnemius	Calf muscle	Knee and ankle	Plantar flexion of ankle Flexion of knee	Calf raises
Soleus	Calf muscle	Ankle	Plantar flexion of ankle	Calf raises
Tibialis anterior	Front of shin	Ankle	Dorsi-flexion of ankle	Toe tapping
Hamstrings	Back of the thigh	Knee and hip	Flexion of knee Extension of hip	Hamstring/ leg curls
Quadriceps	Front of the thigh	Knee and hip	Extension of the knee Flexion of hip	Squats Lunges Deadlift Leg extension
Gluteus maximus	Buttock	Hip	Extension of the hip	Rear leg raises Deadlift Dumbbell lunge squats
Iliopsoas (hip flexor)	Front hip	Hip	Flexion of the hip	Knee lifts
Abductors	Outside of hip and thigh	Hip	Abduction of hip	Side leg raises
Adductors	Inside thigh	Hip	Adduction of hip	Inner thigh raises
Rectus abdominus	Abdominals (front)	Spine	Flexion of the spine	Sit-ups/curl-ups Reverse curls
Erector spinae	Back of spine	Spine	Extension of the spine	Back extensions Deadlift
Obliques	Side of trunk	Spine	Lateral flexion and rotation of the spine	Twisting sit-ups Side bends
Pectorals	Front of the chest	Shoulder	Adduction and horizontal flexion of the shoulder	Press-ups Bench press Bent arm pullover Dumbbell flyes

Table 4.3	Joint/muscle actions (cont.)			
Muscle	Position	Joints crossed	Prime action	Example exercise:
Trapezius	Upper and middle back	Shoulder girdle	Extension of neck Elevation of the shoulder girdle Retraction of the scapula Depression of shoulder girdle	Upright row Shoulder shrugs Prone flyes
Latissimus dorsi	Side of the back	Shoulder	Adduction and extension of the shoulder	Single arm row Bent arm pullover Lat pull down Bent over row
Deltoids	Top of the shoulder	Shoulder	Abduction, flexion and extension of the shoulder	Lateral raise Shoulder press
Biceps	Front of the upper arm	Elbow and shoulder	Flexion of the elbow	Barbell curl Hammer curls Screw curls Concentration curls
Triceps	Back of the upper arm	Elbow and shoulder	Extension of the elbow	Press-ups Tricep dips Lying tricep extension Overhead tricep press Shoulder press Chest press Tricep kick back

Adapted from: Lawrence (2004a).

lift a resistance creating a shortening of the muscle (see Fig. 4.1).

Eccentric: where the two ends of the muscle (origin and insertion) move further apart to lower a resistance creating a lengthening of the muscles (see Fig. 4.2).

Static: where the muscle contracts but stays the same length to hold a position (see Fig. 4.3).

Different sporting activities will demand different types of muscle work. Most require attention to working a muscle through a full range of motion, which requires the specific muscles to shorten

Fig. 4.1 Shoulder press up (lifting phase) = concentric

Fig. 4.2 Shoulder press down (lowering phase) = eccentric

and lengthen (isotonic movements – concentric and eccentric). For some sporting activities static (isometric) muscle work may also need to be included.

Some of the advantages and disadvantages of isotonic and isometric muscle work are listed in Table 4.4.

A further consideration for sport specific training is the type of muscle fibre that will be used. There are different types of fibres within skeletal/voluntary muscle. These include:

- Fast glycolytic fibres (FG) – type IIb
- Fast oxidative glycolytic fibres (FOG) – type IIa
- Slow twitch fibres

The fibres can be differentiated by a number of structural features, which include their preferred energy source (aerobic or anaerobic) and the speed at which they contract and fatigue. The type of muscle fibre recruited (by the nervous system) will be determined by the specific activity. The different types of muscle fibre are described in Table 4.5.

Muscles are usually composed of a

Fig. 4.3	Isometric bars/ring – gymnastics = static

percentage of each type of fibre. This is usually determined genetically. A key factor that influences training potential would be the percentage of each type of fibre an individual possesses. Endurance athletes would most likely have a greater number of slow twitch fibres, whereas power athletes would most likely have a greater number of fast twitch fibres.

A further influencing factor would be the number of type IIa (FOG) fibres as these fibres can respond to training by developing the qualities of either the other two types, depending on the activity being performed. As an average, Sharkey (1990: 273) indicates that the percentage of muscle fibres are:

- 50% slow twitch
- 35% intermediate/FOG
- 15% fast twitch

Neuromuscular connections

All skeletal muscle work is under conscious and voluntary control. It is controlled by the

Table 4.4	Advantages and disadvantages of isotonic and isometric muscle work:	
Isotonic muscle work	Isometric muscle work	
Muscle strengthened through a full range of movement.	Muscle strengthened in specific range of movement	
Related to most sporting activities and daily tasks.	Useful in rehabilitation programmes (strengthens muscle without excessive movement of joint).	
Recruitment of a larger proportion of muscle fibres and motor nerves to work through full range of movement.	Recruitment of specific muscle fibres to hold the position.	
Increased capillarisation – (endurance training)	May cause rise in blood pressure	
	May cause breath holding	

Table 4.5	Muscle fibres		
	Fast twitch Fast glycolytic	Intermediate Fast oxidative glycolytic	Slow twitch
Aerobic capacity	Low	Medium	High
Anaerobic capacity	High	Medium	Low
Size	Larger	Larger	Smaller
Capillary supply	Low	Medium	High
Mitochondria	Low	Medium	High
Contractibility	Very quick	Quick	Slow
Fatigability	Very quick	Quick	Slow
Colour	White	Pink	Red
Intensity	High	Moderate	Low
Duration	Short	Moderate	Long duration
Activities	Power lifting Strength training 1RM – 6/8 RM	Circuit training stations 400 metres	Endurance training Long distance running, swimming etc.

Information sourced from: Sharkey (1990); Bursztyn (1990).

transmission of messages to and from the central nervous system.

Sliding filament theory

When muscles are stimulated to contract, activity takes place deep within the muscle. The small projections/cross-bridges of the myosin filament attach to the actin filaments and pull them inwards – the filaments slide. This causes a shortening of the sarcomere, which in turn creates the shortening of the whole muscle.

Motor unit activation

Messages from the central nervous system are received by motor units.

Larger muscles (quadriceps, biceps) may have single motor units responsible for activating hundreds of muscle fibres. Smaller muscles (eyes, fingers, hands) may have single motor units responsible for activating fewer muscle fibres. The latter would allow for more intricate and delicate movements.

For example:
Playing a musical instrument (piano/guitar) would require more intricate motor unit and muscle control.

Getting out of the chair after playing the musical instrument would require less intricate motor unit and muscle control.

All or none law

When a muscle fibre is stimulated to contract by a motor unit, it will contract fully. Other muscle fibres within the same muscle will not be recruited unless they have also been stimulated to contract. The strength of contraction will depend on the number of motor units and muscle fibres that have been recruited and the speed at which the messages are passed back and forth between the muscle and the CNS. Heavier resistance training and higher intensity movements will generally require the recruitment of more muscle fibres than lower intensity and lower resistance training.

Variables for progressing muscular fitness training

There are a number of variables that can be manipulated to bring about different and specific training gains. These include:

- Repetitions
- Resistance
- Failure
- Rate and range of motion
- Rest (between sets or sessions)
- Sets (single or multiple sets)
- Type and methods (order of exercises, training mode, training systems)

Repetitions

One set of between 8–12 repetitions (whole body) is the guideline for muscular fitness (ACSM: 2000).

Increasing the repetitions will concentrate the training focus on endurance. The longer the duration and higher the repetitions the more the focus will be towards aerobic rather than anaerobic endurance. The specific sporting or training requirements will determine the appropriate repetition range.

Resistance

Increasing the resistance will concentrate the training focus more on strength. The heavier the resistance, the lower the potential for performing repetitions.

Repetition maximum

- 1RM (repetition maximum) is the maximal weight that can be lifted once by a specific muscle/exercise
- 6RM is the maximal weight that can be lifted six times by a specific muscle/exercise
- 8RM is the maximal weight that can be lifted eight times by a specific muscle/exercise
- 15RM is the maximal weight that can be lifted fifteen times by a specific muscle/exercise

1RM can be found directly by selecting an exercise and finding the maximum that can be lifted *once*. There is a risk of injury and a lot of trial and error attached to finding 1RM. Thus, finding a 3RM and then estimating the 1RM offers a slightly safer method. Alternatively, finding a 6–8RM would offer an equivalent to between 70–80% of the 1RM (Bean: 2001: 99) and again offers a safer estimate towards what the 1RM would be.

There are different methods for altering the resistance. These include:

- Increasing the length of the lever being moved – rear leg raise with bent leg will be easier than if the leg is straightened
- Adding body weight to the end of the lever – curl up with hands at side of head will be harder than when hands are on the thighs
- Working against gravity or working across gravity – push ups against a wall will use less gravity than push ups on the floor
- Adding an external resistance – fixed weights, free weights or other equipment described in part 4 will all increase the resistance

Failure

Failure occurs when no more repetitions of the exercise can be performed. Training to failure enables maximal recruitment of motor units (nerves) and thus, greater recruitment of muscle fibres.

There is a point during the concentric phase of all exercises where gravity and leverage make it harder for the resistance to be moved. This sticking point is usually where failure is reached. Some training programmes use partner assistance (assisted lifting) to work through this phase.

Rate and range of motion

The recommendation is for exercise speed to be controlled to promote full range of motion and full use of concentric and eccentric phases – isotonic muscle work.

However, for some sporting events, strength and endurance are needed at specific ranges of motions and also specific speeds of movement. The emphasis of training can be adapted by changing the speed and range of movement.

Working slower on the lifting phase will emphasise concentric muscle work. Working slower on the lowering phase will emphasise eccentric muscle work. The ACSM (2000) guideline is to maintain caution when focusing on eccentric training due to the increased potential for delayed onset muscle soreness (DOMS).

The range of motion can be increased by introducing equipment.

- Incline chest press can be performed on a incline bench
- Incline curl-ups and reverse curls can be performed by using steps with more risers at one end
- Tricep dips can be performed from the edge of a bench or step to increase range of motion
- Squats can be performed from a step to increase the range
- Calf raises can be performed from a step to increase the range

For example:
When working at higher ranges of repetition maximum (ATP-CP), a longer rest time will be required during multiple set training (more than 1 set) to enable ATP-CP to be replenished and prevent the accumulation of lactic acid.

Fig. 4.5 Incline curl-ups and reverse curls can be performed by using steps with more risers at one end

Fig. 4.6 Incline chest press performed on an incline bench

Fig. 4.7 Tricep dips can be performed from edge of bench or step to increase range of motion

Fig. 4.8 Squats can be performed from a step to increase the range

Fig. 4.9	Calf raises can be performed from a step to increase the range

Rest

Rest between sets of exercises will be determined by the intensity of the resistance lifted and the energy system targeted. The higher the resistance/intensity, the longer the rest time will be needed between sets.

Rest between sets

Bean (2001: 105) offers guidelines for resting between sets according to intensity and the training goal (see Table 4.6).

Bompa and Carrera (2005: 73) suggest the following guidelines depending on the type of strength training (Table 4.7).

Rest between training sessions is essential to allow muscle time to recover. The recommendation is for the same muscle to be worked two to three times a week on non-consecutive days. This allows time between sessions for the muscles to be refuelled by replenishing glycogen stores and for any damage to be repaired by new protein.

Individuals training on consecutive days would usually train using a split routine, focusing on specific muscles groups for each workout.

Within a split routine, different muscle groups are worked on different days of the week (in a set order/rotation) with a minimum of two days rest between specific muscle workouts and ideally with two rest days during each rotation, to enable time for recovery. The actual recovery time between workouts is determined by the intensity of the workout, the experience of the individual and their diet. If the muscles still feel sore, the advice would be to continue resting them. It would be counterproductive to train a muscle while it is recovering.

Split routines usually use a wider range of exercises for the specific muscle areas targeted in that session and use more sets of specific exercises (multiple rather than single set). For example, three to five chest exercises, three to five shoulder exercises, three to five tricep exercises. NB: many chest and shoulder exercises also work triceps (chest press, shoulder press). The aim of using a broader variety of exercises is to use more of the muscle fibres in specific movement ranges (incline, decline). It also enables an increased overall volume of training (frequency, intensity, time).

Sets

Single set training is recommended as the general training guideline. Multiple set training can be used for specific training goals and will bring some increased gains. There is much

Table 4.6	Guidelines for rest intervals between sets (Bean. 2001: 105)				
Training goal	Number of sets/exercise	Repetitions	Training weight	Rest intervals	Training tempo Lift: lower
Maximum strength	2–6	<6	Heavy (>85%)	2–5 Min	1:2
Power	3–5	1–5	Heavy 75–85%	2–5 Min	Explosive:1
Muscle size	3–6	6–12	67–85%	90 sec	2:3
Muscle Endurance	2–3	>12	Lower than 67%	<30 sec	2:3

Table 4.7	Guidelines for rest intervals between sets (Bompa and Carrera. 2005: 73)		
Load %	Speed	Rest interval duration in minutes	Application
> 105 eccentric	Slow	4–5	Maximum strength and muscle tone
80–100	Slow to medium	3–5	Maximum strength and muscle tone
60–80	Slow to medium	2	Muscle hypertrophy
50–80	Fast	4–5	Power
30–50	Slow to medium	1–2	Muscular endurance

controversy related to the use of single or multiple set training (ACSM: 2000, Fleck & Kraemer: 1987, Bean: 2001).

Single set advocates tend to ignore the use of warm-up sets within their programming, which could be considered as working sets. Advocates of multiple set training include the warm-up sets as part of the programming for a working set (Bean: 2001). The main aim and consideration is for overload to be achieved. The rest period between sets of exercises will be determined by the resistance lifted. The heavier the resistance the longer the rest/recovery periods between sets.

Table 4.8	Split routine example 1
Monday	Back and biceps
Tuesday	Chest, shoulder and triceps
Wednesday	Legs, hips and abdominals
Thursday	Rest
Friday	Back and biceps
Saturday	Chest, shoulder and triceps
Sunday	Legs, hips and abdominals
Monday	Rest

Table 4.9	Split routine example 2
Monday	Back and chest
Tuesday	Shoulder and arms
Wednesday	Rest
Thursday	Legs and abdominals
Friday	Rest
Saturday	Chest and back
Sunday	Rest
Monday	Legs and abdominals

From: Bean (2001: 117)

Type and methods

Order of exercises.

Compound – Isolation

Compound exercises involve more than one larger muscle group to perform an exercise. They have the advantage that more muscles can be targeted in a specific exercise.

Compound exercise example:
Chest press (compound) – chest, triceps and anterior deltoid
Lat pull down (compound) – back, biceps and brachialis
Leg press (compound) – quadriceps, hamstrings, gluteals

Isolation exercises involve smaller and single muscle work. They have the advantage of targeting a specific muscle area, which can be useful for correcting muscle imbalance.

Isolation exercise example:
Tricep extension (isolation) – triceps
Bicep curl (isolation) – biceps
Back extension (isolation) – erector spinae
Leg extension (isolation) – quadriceps

Large to small muscles

It is usual to work using the larger muscles (compound exercises) first and then the smaller muscles (isolation).

Compound exercises may demand greater exercise technique because more than one joint action is involved (for example: chest press – elbow flexion/extension and shoulder – horizontal flexion/extension).

Some advanced training programmes (pre-exhaust) use isolation exercises before compound exercises with the intention of exhausting the larger muscle group first.

For example:

- Chest flyers (isolation)
- Chest press (compound)

Example:
Upper body – Lower body alternating exercises
0 Squats
1 Shoulder press
2 Lunges
3 Chest press
4 Step ups
5 Bent over row

Upper and lower body

Performing an upper body exercise and then a lower body exercise is common in circuit training programmes.

This method enables one area to rest while another is active and can assist with maintaining the intensity to achieve cardiovascular as well as muscular fitness goals. The disadvantage is that the muscular focus is reduced and less strength gains may be achieved in this specific component. This approach is less appropriate for advanced trainers who want to focus on building strength or muscle bulk.

Training mode: free weights, fixed weights, body weight

There is a range of different mediums for training strength and endurance. These include: working with free weights, working with fixed resistance machines and working with body weight. Other equipment, as described in part 4, can also be used for advanced and hardcore circuit training.

Some considerations and benefits of working with free weight, fixed resistance machines and body weight are discussed in Table 4.10.

Resistance training systems

There is a range of different resistance training methods that can be used within the design of a circuit training programme. Some are discussed in the following paragraphs.

Supersets

There are two methods of super setting.

Method one

Involves working the agonist and then the antagonist muscle

Example:
- Barbell bicep curl followed by tricep dips
- Press-ups followed by bent over row
- Leg extension followed by leg curls

Method two

Involves performing a number of different exercises for the same muscle group in succession. This is sometimes referred to as compound/tri/giant sets.

Example:
- Tricep dips
- Tricep kick backs
- Narrow arm press-ups
- Overhead tricep press

Both methods involve continuous movement from one exercise to the next without rest. Usually 8–10 repetitions of each exercise are performed. Body builders use this programme for muscular hypertrophy and growth (Fleck and Kraemer. 1987: 91)

The benefit of the super setting method is that it enables continuous muscle work, which potentially reduces the work time. The

Table 4.10	Considerations and benefits of training with free weights, fixed resistance machines and body weight		
	Free weights This includes some of the external resistance equipment described in this book (kettlebells, medicine balls, tyres etc).	**Fixed weights**	**Body weight**
Range of motion	Enables full potential range of motion.	Range of motion is fixed by machine. Some machines do not accommodate different body types and lever lengths.	Enables full potential range of motion.
Fixator muscles	Greater fixation of surrounding muscles to control movement.	Fixed machine provides stability to movement. There is potentially less stability work required of other muscles.	Greater fixation of surrounding muscles to control movement.
Resistance	Weight can be added in smaller increments, e.g. dumbbell increments: 2kg, 3kg, 5kg, 7.5kg, 10kg, 12kg, 15kg, 18kg, 20kg etc. Barbell plates also come in a range of sizes to enable weight to be added progressively.	Some fixed machines have larger resistance increment changes, e.g. 5kg to 10kg	Other variables (range of motion, leverage, gravity) need to be manipulated to add resistance. There is a limit to the resistance that can be lifted.
Safety	Requires lots of control/skill to handle free weights. Heavy resistances require use of spotter/racks. Weights need to be lifted into position safely (dead lift, clean or use of rack). If weight is dropped the lifter could be injured (particularly for bench lifts).	Fixed machine limits range of motion. No spotter required. If weight is dropped, no risk of it landing on the lifter.	Provided correct technique and speed of movement is used, potentially less risk of injury.

NB: there is an element of risk attached to all forms of training. The key is to be aware of the potential risks and seek to minimise them.

disadvantage would be that it demands a good base level of fitness.

These methods can be easily used in a circuit training programme. Stations can be grouped to work agonist and antagonist muscles (method one) or they can be grouped to work agonist muscle in different ways (method two).

Super setting can be a fabulous way of adding challenge within a circuit training programme.

Pre- and post-exhaust

Pre-exhaust involves working and tiring/exhausting stronger muscle groups used in a compound exercise and then performing the compound exercise so that the weaker muscle completes most of the work.

For example:
Chest flyes – to pre-exhaust the pectorals
Chest press – for triceps to work harder

A variation to pre-exhaust is pre- and post-exhaust
For example:
Chest flyes – to pre-exhaust the pectorals
Chest press – for triceps to work harder
Chest flyes – to post-exhaust the pectorals
Both methods can be used in circuit training by grouping exercise stations together.

Example one:
Leg extensions with medicine ball to pre-exhaust quadriceps
Barbell squats – for gluteals to work harder

Example two:
Tricep dips – to pre-exhaust triceps
Press-ups – for pectorals to work harder
Tricep dips – to post-exhaust triceps

The above are simple examples of the method being applied to practice. There are many vari-

Fig. 4.10 Press-up

Fig. 4.11 Tricep dip

ations and a number of different exercises that can be used to achieve this approach.

Pyramids

Pyramid or triangle programmes are used predominantly by power lifters.

A full pyramid would start with a low resistance for 10–12 repetitions. This would develop over a number of sets by performing fewer repetitions with a heavier resistance up to a level where a maximum of 1 repetition would be performed. The process would then be reversed altering the same variables to lower the resistance and increase the repetitions.

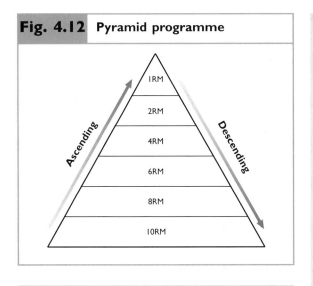

Fig. 4.12 Pyramid programme

(pyramid levels, top to bottom: 1RM, 2RM, 4RM, 6RM, 8RM, 10RM; Ascending arrow on left side, Descending arrow on right side)

Example:
10 repetitions – increase resistance
8 repetitions – increase resistance
4 repetitions – increase resistance
2 repetitions – increase resistance
1 repetition – increase resistance
2 repetitions – decrease resistance
4 repetitions – decrease resistance
8 repetitions – decrease resistance
10 repetitions – decrease resistance

This method can be varied by just performing the ascending first half of the pyramid/triangle (light to heavy) or performing the descending half of the pyramid/triangle (heavy to light).

Circuit example:
One station
10 repetitions – box press-ups
8 repetitions – $^3/_4$ press-ups
4 repetitions – full press-up single count
2 repetitions – slow full press-ups (2 up, 2 down count)
1 repetition – super slow full press-ups (4 up and 4 down count)
2 repetitions – decrease resistance (2 up, 2 down count)
4 repetitions – decrease resistance – full press-up single count
8 repetitions – decrease resistance – $^3/_4$ press-up
10 repetitions – decrease resistance – box press-ups

NB: the variations listed above offer one example. Pyramids can also be applied using the same positions but changing the speed of movement.

Summary

The key points relating to muscular fitness training discussed in this chapter include:

- Muscular strength is improved by training with higher resistance and lower repetitions
- Muscular endurance is improved by training with lower resistance and higher repetitions
- Muscular fitness training offers a procedure to optimize everyday functioning
- The frequency, intensity, time and type of training will vary depending on the fitness goals
- A variety of resistance training systems can be used: single sets, super sets, giant sets, pyramid, split programmes etc
- Varying the repetitions, resistance, rate, range of motion will alter the effect of the training.

MOTOR FITNESS

5

Motor fitness is a skill-related component of fitness and includes the following sub-components:

- Reaction time
- Agility
- Balance
- Speed
- Co-ordination
- Power

Managing our body weight, manoeuvring our centre of gravity, co-ordinating movements of the arms and legs, moving at different speeds, changing direction, throwing, jumping, catching, punching/boxing, dancing all require motor fitness.

If we want to improve our motor fitness, we must specifically and repeatedly train the aspect we wish to improve. Training to improve each sub-component of motor fitness will require different and specific activities. For example, to perform a quick, co-ordinated sequence of movements requires practice of the specific movements that make up that sequence. Improving balance will require different training than that required to develop power etc. The application of the skill (the sport or athletic event) will also demand a specific training focus.

Motor fitness requires the effective transmission of messages and responses between the central nervous system (the brain and spinal cord) and the peripheral nervous system (sensory and motor). The peripheral system collects information via the sensory system, the CNS receives and processes this information and sends an appropriate response via the motor system, which initiates the appropriate response.

A brief overview of the structure and function of the nervous system is described later in this chapter. A description of the sub-components of motor fitness is provided first.

Power

Power is often referred to as a combination of strength and speed of movement. Power is a key component in athletic field events such as javelin, long jump, high jump, shot putt, discuss etc. To improve power for specific events requires training of the specific movement that is equivalent to the performance/competition movement speed. The onset of fatigue is high in power training, thus appropriate rest needs to be planned between work intervals.

Scholich (1999: 12) suggests the following training regime: sub maximal but explosive force or strength application (75–90%) with recovery periods of 90–240 seconds to ensure optimum recovery.

Co-ordination

Co-ordination is the skill required to manage complex movements efficiently and effectively. It is demanded at different levels during a range of sports and activities. Co-ordination is needed to:

- Tie shoe laces
- Drive a car
- Walk in a straight line

- Perform a dance or gymnastics routine
- Dribble a football around cones
- Hit a ball with a racquet

Skilled performers make activities appear easy, when often there is a series of well timed and highly co-ordinated processes occurring to achieve this look. Skilled performance demands practice and repetition to refine movement patterns to high levels.

When learning a new skill, a lot of conscious control is needed initially. As skill develops, the process becomes more automatic, almost unconscious. For example, when learning to drive the mind is initially racing between controlling the gear stick, foot pedals, looking at the road etc. Eventually, all of these are managed without such conscious thought.

Balance

Balance of the body is managed when the centre of gravity (pelvic region between the hips and sacrum) is positioned centrally over the base of support (the feet when standing, the buttocks when sitting). The wider the base of the support and the lower the centre of gravity, the easier it will be to balance the body.

For example:
Lying on the floor will offer more balance than standing – when lying the base of support is wide and the centre of gravity is low.

Standing with a wide foot stance and bent knees to widen the base of support and to lower the centre of gravity will offer more balance than standing with feet together where the base of support would be narrower and the centre of gravity higher (see figs. 5.1 and 5.2).

Standing with feet together will offer more balance than standing on the ball of the foot on one leg where the base of support would be smaller and the centre of gravity higher.

Fig. 5.1 Sumo wrestler in wide stance

Fig. 5.2 Ballet dancer standing on one leg

Balance is maintained through use of the senses (eyes, ears and receptors in the muscles, tendons and joints). Removing any of these senses (closing eyes, covering ears) will reduce incoming input that assists the brain with monitoring balance.

Agility

Agility is defined by Davis, Kimmet and Auty (1986: 42) as: 'the ability to handle the body quickly and precisely to change direction accurately while moving quickly'.

Agility is an essential component during games (catching, passing, dribbling, dodging etc). It can be developed by performing activities that mimic those required for the sporting event, such as dribbling a ball around cones, shuttling from one side of a badminton court to the other.

Speed

Speed is defined by Davis, Kimmet and Auty (1986: 43) as: 'the ability to put the body or parts into motion quickly'. Speed is measured by the duration it takes to complete the specific task, which is determined by reaction time and movement time.

Speed of movement is necessary for a range of track events (sprinting) and sporting games (football etc).

To develop speed it is necessary to work on both reaction time and movement time.

Reaction time

Reaction time is described by Davis, Kimmet and Auty (1986) as the: 'amount of time between a stimulus and the first movement initiated in response to it'.

Examples of reaction time in action include: the speed at which a sprinter moves from the blocks after hearing the start gun fire; the speed at which a footballer receives and manages the ball after a pass from another player; the speed

at which a cricketer can hit the ball after the bowler releases the ball.

Movement time

Movement time is the time taken to complete a movement after it has been initiated such as completing the 100 metre sprint.

Reaction time, movement time and speed of movement are essential for sports and provide the edge between winning and losing. The faster the reaction time and movement speed, the longer the potential for processing information, thereby giving extra time to monitor other stimuli (position of other players etc).

The great boxer Muhammad Ali is reported as being able to move his fist 40 centimetres in 0.06 seconds! (Davis, Kimmet, Auty. 1986: 295).

Brief overview of the nervous system

The nervous system is the control centre of the body. There are two main sections:

- The central nervous system (CNS) comprising the brain and spinal cord. This is the main processing centre
- The peripheral nervous system (PNS) comprising the nerves that branch away from the spinal cord to the rest of the body. This includes the nerves to the muscles that are used to perform sports and activity

The peripheral system consists of nerves that carry information:

- To the CNS and away from the body – sensory nerves or afferents
- From the CNS and to the body – motor nerves or efferents

Fig. 5.3 — The central nervous system (CNS)

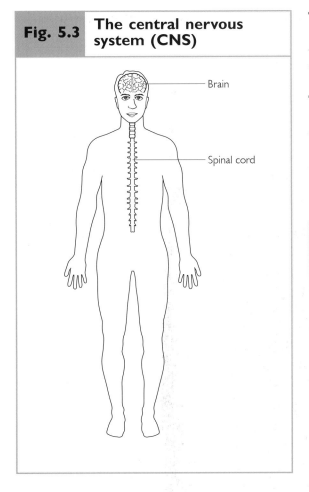

Brain

Spinal cord

There are two divisions of the peripheral systems:

- The somatic system
- The autonomic system

The somatic system

- Responds to stimuli outside the body (external senses)
- Controls the movement of the skeletal muscles
- Conscious/voluntary control

For example:
The eyes of a cricketer will observe when the bowler releases the ball. The speed at which the information is processed and the appropriate responses relayed back via the motor nerves will determine how the cricketer hits the ball.

The ears of a sprinter will hear the fire of the start gun. The speed at which the information is transmitted and processed will determine the speed at which they react and leave the start block.

Another example would be through the sense of smell (nose) – if we detect an obnoxious odour, the quicker the information is processed the quicker we can move away from the smell.

The examples listed previously (cricket and sprinting) would be elicited via the somatic system.

The autonomic system

- Responds to stimuli inside the body (internal senses)
- Manages processes such as heart rate, digestion
- Unconscious control – involuntary

For example:
We respond to hunger by eating (conscious – somatic)
Digestion of food occurs without conscious control (unconscious – autonomic)

Sensory nerves/afferents include the eyes, nose, ears, skin, mouth (the senses) and also proprioceptors (stretch reflex, Golgi tendon organ). The nerves collect information from the body and send this to the brain for processing.

Motor nerves/efferents include the connections, which manage muscle work (motor unit recruitment, all or none law etc). The nerves which relay messages from the brain and spinal cord (CNS) to evoke action in the body.

Summary

The key points discussed in this chapter include a description and guidelines for improving each component of motor fitness:

- Power
- Speed
- Balance
- Agility
- Reaction time
- Co-ordination

Motor fitness is a skill-related component of fitness.

The nervous system is the control centre of the body.

The central nervous system (CNS) consists of the brain and spinal cord.

The peripheral nervous system (PNS) consists of the nerves that branch away from the spinal cord to the rest of the body and includes the nerves to the muscles that are used to perform sports and activity.

SESSION STRUCTURE AND PROGRAMME DESIGN

All sessions must be structured carefully and progressively to achieve safe and effective work-outs and training programmes, appropriate for the exercisers. All exercisers will need to be warmed up thoroughly before work and cooled down thoroughly after work. This section reviews session structure and programming.

Chapter 6 reviews the aims and structure of the warm-up, illustrates and explains a range of dynamic stretches that may be included within the warm-up for advanced exercisers.

Chapter 7 reviews the aims and structure of the cool down and illustrates and explains a range of partner stretches that can be used with advanced exercisers.

Chapter 8 reviews seasonal training aims and goals for sports specific training.

Chapter 9 introduces periodisation.

Overview of session structure

Warm-up	Main workout	Cool down
Mobility exercises for joints to be used in main session. Pulse raising exercises Stretching exercises	Circuit stations Cardiovascular and/or muscular focus (exercise stations).	Pulse lowering exercises Stretching exercises for muscles worked in main session.
Duration and intensity – variables	**Duration and intensity – variables**	**Duration and intensity – variables**
Duration and intensity of warm-up. Number of stretches Type of stretching – static or dynamic. Intensity of stretch positions.	Work time per station Rest time per station Type of rest – active or passive Number of stations Type of stations – cardiovascular, muscular fitness, skill related. Equipment used Intensity of exercises selected Number of circuits Layout of circuit	Duration and intensity of pulse lowering exercises. Intensity of stretch positions. Length of hold (maintain/develop). Number of muscle groups stretched. Number of stretches for each muscle group.

ADVANCED WARM-UPS

6

The warm-up is the preparation of the body for exercise. All exercisers need an appropriately structured warm-up containing safe and effective exercises to prepare the body systems for the main workout.

The warm-up should contain:
- Mobility exercises for all the main joints that will move in the session. Mobility exercises warm the joints and stimulate the release of synovial fluid into the joint capsule. Synovial fluid lubricates the joints and eases joint movement. Mobility exercises will also engage the muscles around the joint providing the preliminary range of motion work necessary to release tension in the muscles, which again assists with joint movement. Table 6.1 lists a range of mobility exercises.
- Pulse raising exercises to gradually elevate the heart rate and increase the flow of blood and oxygen to the working muscles so that they are able to work effectively using the aerobic energy system. Table 6.1 lists a range of pulse raising exercises.
- Stretching exercises to lengthen the muscles further and take them to a range at which they may be extended within the main workout. A range of dynamic stretches are illustrated and explained at the end of this chapter.

There is continued debate regarding the inclusion or exclusion of a preparatory stretch phase within the warm-up. Most debates recommend that some range of motion activity (mobility and or dynamic stretching that takes the muscle/joint through its full range of motion) is included within the preparatory phase as a way of lengthening the muscle to a working range.

Proponents against static stretching quite accurately argue that muscle warmth can be lost when performing static stretching for too long, thus losing the benefits of the warm-up.

However, the ACSM (2006) recommend the use of static stretching with general populations and thus to follow this guideline and maintain the inclusion of stretching within the warm-up, it is suggested that other muscle warming activities (pulse raising) can be performed between stretches as a way of maintaining the effects of the warm-up.

Alternatively, experienced exercisers with good levels of body awareness may include dynamic and a range of motion stretches within the warm-up phase. The emphasis for dynamic stretching should be on lengthening the muscle to the end of the existing/current range of motion and then moving in a controlled fashion out of the stretch position. This process is repeated for a number of repetitions – moving with control into and out of the stretch position, without over stretching or exceeding the range of motion.

Our personal recommendation is that instructors review the purpose of stretching and the underpinning knowledge relating to flexibility and stretch training. All too often, static pre-stretching is omitted from the warm-up *but* there is no, or very limited range of motion preparation included in the warm-up either! If

Table 6.1 Mobility and pulse raising exercises

Mobility exercises	Pulse raising exercises
Shoulder lifts	Walking
Shoulder rolls	Marching
Single arm circles	Squatting
Double arm circles	Lunging
Lateral raises	Box steps
Chest press arm action	Step-ups
Shoulder press arm action	Side stepping
Pull down arm action	Calf raises with arm reaching
Side bends	Gentle jogging
Trunk rotations	Back lunge steps
Pelvic circles	Side lunge steps
Leg curls	
Knee raises	NB: any variation of arm movement
Foot peddles	can be combined with the above, for
Calf raises	example:
	Bicep curls
	Shoulder press
	Lat pull down action
	Chest press action
	Tricep kick back action
	Pec dec action
Key considerations	**Key considerations**
Start with smaller range mobility exercises and progressively build range of motion.	Integrate pulse raising with mobility and exercise building intensity progressively by making movements deeper, travelling further, combine with arm movements and adding light impact.
	When warm integrate larger range mobility work, dynamic and range of motion stretching or static stretching if preferred.
NB: if static stretching is used, ensure pulse raising is maintained to keep the body warm.	

NB: most of these exercises are illustrated and explained in *Circuit Training: A Complete Guide to Planning and Instructing* (Lawrence & Hope: 2007).

static pre-stretching is not included, it is essential that there is some range of motion activity to ensure the joints and muscles are adequately prepared and lengthened to their working range.

In addition, post-workout stretching should contain stretching for muscles that have been worked in the session and for muscles that lack flexibility. It would be less relevant to include a stretch for the posterior deltoid stretch if this was not used within the session (for example, a session focusing on training the chest or the legs). A further consideration would be that since many people may have rounded shoulders, the relevance of the posterior deltoid stretch could be further questioned! The question being 'Would this muscle not benefit more from being strengthened?'

We recommend that both pre- and post-stretching should focus on the muscles used within the session.

An example of some dynamic stretches that may be included as part of the warm-up are described and illustrated. It is advisable that the body is warm before moving to greater ranges of motion. The purpose of all these stretches is to stretch the specific muscle with a dynamic range of motion.

Dynamic stretches/larger range mobility exercises

Exercise 6.1 — **Dynamic adductors stretch – fixed side lunges**

Purpose

This exercise is used to dynamically stretch the adductors.

Starting position and instructions

- Legs positioned approximately two shoulder widths apart
- Lunge from right side to left side in a controlled manner

Coaching points

- Knees move in line with the toes, without overshooting the ankles
- The movement can start small and progress to a larger range of motion by bending deeper
- The pelvis stays facing forward
- Abdominals engaged
- An upright posture is maintained throughout
- The shoulders relaxed and down away from the ears

Progressions/adaptations

- Static stretch by holding on each side for the less skilled
- Increase range of motion for individuals with greater flexibility

Exercise 6.2	**Fixed leg forward lunge**

Purpose

This exercise is a dynamic hip flexor stretch.

Starting position and instructions

- Stride one leg forward of the other (a large stride length)
- Lower the body weight down towards the floor, bending both knees

Coaching points

- Front knee moves in line with toes, without overshooting the ankles
- Back knee sinks towards the floor without hitting the floor
- The movement can start smaller and progress to a larger range of motion by bending deeper
- The pelvis stays facing forward

- Abdominals engaged
- An upright posture is maintained throughout
- The shoulders relaxed and down away from the ears
- Repeat with other leg forward

Progressions/adaptations

- Static stretch – holding position for the less skilled
- Increase range of motion by bending further into stretch for more skilled/flexible individuals

Exercise 6.3	**Dynamic erector spinae stretch – hollow and rounding**

Purpose

This exercise is a dynamic erector spinae stretch.

Starting position and instructions

- Feet a little wider than hip width apart, with knees bent

- Bend forward at hip and rest hands on the thighs
- Round the lower spine by tucking the tail bone under
- Reverse and hollow the lower back lifting the tail bone up

Coaching points

- Move with control and to a comfortable range of motion
- The pelvis stays square and weight is spread between both legs
- Knees over ankle and not rolling inwards
- Abdominals engaged
- The shoulders relaxed and down away from the ears

Progression/adaptation

- The movement can start smaller and progress to a larger range of motion by rounding and hollowing further, but within a comfortable range and without jerking the movement

Exercise 6.4	Dynamic calf stretch – fixed leg rocking

Purpose

This exercise is a dynamic calf stretch.

Starting position and instructions

- Stride one leg forward of the other (a large stride length)
- The front leg should have a slight bend and the back leg should be straight
- Keep the weight on the front leg
- Transfer the weight forward by rocking slightly and lifting the back heel
- Lower the back heel down to the floor, letting the weight shift backwards
- Perform for a few repetitions

Coaching points

- The pelvis stays facing forward
- Abdominals engaged
- An upright posture is maintained throughout
- The shoulders relaxed and down away from the ears
- Repeat with other leg forward

Progression/adaptation

- The movement can start small and progress to a larger range of motion moving the back leg a little further backwards through the movement, if comfortable

Exercise 6.5	Reverse pec dec

Purpose

This exercise is a dynamic chest stretch.

Starting position and instructions

- Can be performed while walking or marching
- Raise the hands to a pec dec position (option 2 – place hands at the side of the head)
- Shoulder relaxed and down away from the ear
- Squeeze the arms backwards, drawing the shoulders down to lengthen the chest muscles and then return to start position, repeat for a few repetitions

Coaching points

- The pelvis stays facing forward
- Abdominals engaged
- An upright posture is maintained throughout

Progression/adaptation

- The movement can start small and progress to a larger range of motion by moving the elbows further back
- Static chest stretch for less skilled
- Increase range of motion for individuals with greater flexibility

Exercise 6.6	Dynamic tricep stretch

Purpose

This exercise is a dynamic tricep stretch.

Starting position and instructions

- Can be performed while walking or marching
- Raise one hand over the head, shoulder relaxed and down away from the ear
- Lower the hand to touch just below the base of the neck at the back of the body and raise back to straight position, repeat for a few repetitions

Coaching points

- The pelvis stays facing forward, abdominals engaged
- An upright posture is maintained throughout
- Repeat with other arm

Progression/adaptations

- Static stretch to modify
- The movement can start smaller and progress to a larger range of motion by moving the hand lower

Exercise 6.7	**Dynamic quadriceps stretch**

Purpose

This exercise is a dynamic quadriceps stretch.

Starting position and instructions

- Legs positioned approximately a shoulder width and a half apart
- Take the weight on one leg and raise the opposite heel to the buttocks
- Repeat by transferring weight to the other leg

Coaching points

- Knees move in line with the toes
- The pelvis stays facing forward
- Abdominals engaged
- An upright posture is maintained throughout
- The shoulders relaxed and down away from the ears
- Perform for a few repetitions

Progressions/adaptations

- The movement can start smaller and progress to a larger range of motion by raising the heels closer towards the buttocks
- Static quadriceps to modify for less flexible individuals
- Increase range of motion by raising heels closer to buttocks

Exercise 6.8	Static hamstring stretch

Purpose

This exercise is a static hamstring stretch.

Starting position and instructions

- Step one leg forward of the other and bend the back knee
- Keeping the weight on the back leg, straighten the front leg without locking the knee
- Lean forward at the hip and aim to place the hands on the bent knee

Coaching points

- The pelvis is square
- Abdominals engaged
- The length of the spine is maintained throughout
- The shoulders relaxed and down away from the ears

NB: a static stretch is offered rather than a dynamic stretch for the hamstrings. This is because it may be safer to lengthen the muscle in this way, rather than moving into and out of the stretch which may be controversial for the spine/back. A further consideration is that the hamstrings are frequently a tighter and some-times shortened muscle group – a static stretch will potentially enable more effective stretching.

Summary

The key points discussed in this chapter include:
- The warm-up needs to contain exercises that mobilise the joints, warm the muscles, elevate the heart rate, increase the blood circulation and lengthen the muscles
- Mobility exercises can be performed to prepare the joints and release synovial fluid
- Pulse raising exercises can be performed to elevate the heart rate
- Static or dynamic stretches can be included to lengthen the muscles
- There will need to be some skill rehearsal in the warm-up to prepare for the main workout.

ADVANCED COOL DOWNS

7

The cool down returns the body to a pre-exercise state. All exercisers need an appropriately structured cool down containing safe and effective exercises to return the body systems back to a comfortable state.

The cool down should contain:

- Pulse lowering exercises to reduce the working heart rate and promote the circulation of blood back to the heart and the brain (venous return). Pulse lowering reduces the risk of blood pooling in the lower extremi-

ties, which can lead to dizziness and fainting. Pulse lowering will also reduce the stress on the cardiovascular and respiratory systems.
- Stretching exercises to lengthen the muscles (the muscles that have been worked on through the session and/or the muscles that need a flexibility focus), maintaining and/or developing/extending the range of movement of specific muscles.

A range of stretches that can be assisted by a partner are now listed. It is essential that partner stretches are only used with advanced individuals.

Pulse lowering exercises	Stretching exercises
The intensity at which exercises are performed should begin at the level of the main circuit and progressively decrease by reducing the impact, the depth of bending movements and reducing the travel and speed of movements.	Stretches should be included for all major muscle groups. Static stretches are recommended. The use of partner stretching is acceptable for advanced groups with high levels of body awareness.
Jogging	Hamstrings
Jump jacks	Adductors
Squatting	Pectorals
Forward lunges	Hip flexor
Backward lunges	Triceps
Side lunges	Quadriceps
Skipping	Obliques
Walking	Latissimus dorsi
Marching	Gastrocnemius
Leg curls	Soleus
Knee raises	

There must be effective communication between stretching partners to ensure a safe range of motion is achieved.

The purpose of each of these stretches is to stretch specific muscles in a way only achievale with your partner's cooperation.

Cool down exercises

| **Exercise 7.1** | **Partner assisted lying hamstring stretch** |

Purpose

To stretch the hamstring.

The stretcher

Starting position and instructions

- Lies on their back with knees bent and feet firmly on the floor at hip width

Coaching points

- Neutral pelvic alignment with a natural curve of the spine
- Engage abdominals
- Straighten one leg and raise actively towards the body
- Keeping head and shoulders on the floor knee is unlocked but fully extended

The partner

Starting position and instructions

- Positions themselves at the side of the stretcher in a kneeling lunge position
- Supports the weight of the stretcher's leg against their body

Coaching points

- When comfortable, the stretcher tells their partner to ease the leg a little closer towards the stretcher's body
- The movement and action should be flexion at the hip, without excessive flexion of the spine
- The position can be held for a comfortable duration

Progression/adaptation

- PNF stretching techniques can be used (see chapter 3 page 29)

| **Exercise 7.2** | **Partner assisted seated pectoral stretch** |

Purpose

To stretch the pectoral muscles.

The stretcher

Starting position and instructions

- Sits upright on the floor (or on a step) with spine lengthened and abdominals engaged
- Takes their hands backwards, drawing the shoulder blades downwards, till a mild tension is felt at the front of the chest

The partner

Starting position and instructions

- Takes hold of the stretcher's arms above the elbows
- The stretcher relaxes their arms, so the weight of the arms is supported by their partner

Coaching points

- When comfortable, the stretcher tells the partner to ease the arms a little further back and down to increase the stretch on the chest/pectorals
- The stretcher should maintain correct alignment of their spine
- The position can be held for a comfortable duration

Progression/adaptation

- PNF stretching techniques can be used (see chapter 3 page 29)

Exercise 7.3	**Partner assisted adductor stretch**

Purpose

To stretch the adductors.

The stretcher

Starting position and instructions

- Sits on the floor with the legs straddled out to the side
- Their hands can be placed on the floor behind the back to help maintain an upright posture

Coaching points

- The abdominals should be lightly engaged

The partner

Starting position and instructions

- Sitting facing the stretcher with an upright posture
- Places their feet inside the stretcher's leg, above the knee – not on the knee joint

Coaching points

- When comfortable, the stretcher communicates with the partner to ease the legs a

little wider to increase the stretch – mild tension
- The position can be held for a comfortable duration

Progression/adaptation

- PNF stretching techniques can be used (see chapter 3 page 29)

| Exercise 7.4 | Partner assisted erector spinae stretch |

Purpose

To stretch the spine.

The stretcher

Starting position and instructions

- Sitting on step with feet firmly on the floor and the knees bent
- Lean forward over the thighs to lengthen the spine, with weight resting on the thighs

Coaching points

- The shoulders should be relaxed and down, away from the ears

The partner

Starting position and instructions

- Stands side on to the stretcher
- Places one hand on the stretcher's sacrum, fingers pointing down towards buttocks
- Places the other hand between the stretcher's shoulder blades

Coaching points

- When comfortable, the stretcher communicates with the partner and the partner adds a gentle pressure through both hands (pushing the hands further apart), to lengthen the spine further
- The position can be held for a comfortable duration

Progression/adaptation

- PNF stretching techniques can be used (see chapter 3 page 29).

Summary

The key points discussed in this chapter include:

- The cool down returns the body to a pre-exercise state
- The cool down should include exercises to lower the heart rate and lengthen the muscles
- Static maintenance stretches will maintain existing levels of flexibility
- Static developmental stretches can be integrated to improve flexibility
- Partner stretches can be used with advanced groups.

SEASONAL TRAINING

8

Most sports and athletic events are seasonal. The training year would include:

• Pre-season training
• Main season training
• Post-season training

Pre-season – developing base fitness levels in preparation for competition

The primary aim of pre-season training is to develop a base level of fitness in all components of physical fitness discussed in part one. Once a base level of fitness is achieved, the sports person can then work towards developing the specific components of fitness and motor skills that will enhance their performance in their chosen sporting event. These will vary depending on the sporting activity, but may include one or more of the components of physical fitness and motor fitness discussed in part one. Pre-season training durations can vary from between six to eight weeks to three months.

Main season – maintenance of fitness and fine-tuning of skills

During main season training, the sports person will need to develop and enhance their sports specific skills while maintaining existing fitness and managing competitions

There is always a risk of injury during competitive seasons. Injury can leave the sports person feeling frustrated and desperate to get back to their training regime, to restore their fitness level. However, any injury sustained would need to be fully rehabilitated before returning to competition. Specific injury rehabilitation programmes should be developed with the guidance of a physiotherapist.

An alternative exercise option to assist recovery from injury is exercise/circuit training in water. Water supports the body so it is possible for the sports person to continue training in water during their recovering period. Fitness can be developed and skills enhanced without placing the injured joint or muscle under too much stress. The use of a buoyancy belt, even for shallow water workouts, will provide further support to the body and reduce the risk of causing further trauma. In addition, the pressure of the water may help to reduce the swelling and reduce the pain at the injured site (Lawrence: 2004a).

Post-season – active recovery period

The primary aim of post-season training is to provide an active recovery period for the body. The body will need time to recuperate and recover from the stresses of competition.

The programme of activities provided at this stage should be sufficient to maintain a reasonable level of physical fitness. However, they should be less demanding. A greater emphasis can be placed on mobility and stretching activities and activities to promote skill development.

Planning sports specific circuits

Sport specific training should follow the same session structure as any other session.

1 There should be adequate warm-up that prepares the body specifically for the main workout
2 The main workout needs to give specific attention to:
   • Replicating the activities performed maximally throughout the sport (joint actions, energy systems, skills etc)
   • Counter-balancing the strength of the opposing muscles to those used maximally throughout the activity

3 The cool down will need to enable the body to cool off and recover. A consideration could be given to developing flexibility in muscles that are required to lengthen quickly when their opposing muscle is contracted quite forcefully. For example, during football the kicking action requires the quadriceps to contract forcefully. The opposing muscle group, the hamstrings, need to be sufficiently flexible to relax and lengthen. If they are not, there is a greater risk of the muscle tearing.

NB: a range of ideas for planning sports specific circuits were provided in *Circuit Training: A Complete Guide to Planning and Instructing*.

Summary

The key points discussed in this chapter include considerations for:
• Pre-season – developing base fitness level in preparation for competition
• Main season – maintenance of fitness and fine tuning of skills
• Post-season – active recovery period.

PERIODISATION

9

Periodisation is a method of manipulating the structure of a training programme to balance out the volume (FITT) and prevent over training (Hayes: 1995). It has consistently been used by track and field athletes and sports people to maximise gains in strength, endurance, power, cardiovascular fitness and achieve peak performance for specific athletic or sporting events. It can also be used as a training method for general populations.

The benefits of a periodised training programme are that it offers:

- A logical and structured approach to progressive training
- A purpose to training with specific goals and targets
- A staged approach for monitoring progression and achievement of goals (short term, medium term and long term)
- A way of assisting psychological motivation
- A method for reducing the potential risk of over-training and subsequent injury
- A way of promoting adherence
- A way of preventing boredom
- A method for avoiding plateaus within training

Periodised training programmes are divided into blocks or layers of training. These include:

- Macrocycles
- Mesocycles
- Microcycles

Macrocycles

A macrocycle represents the longest block of a training programme. The aim of the macrocycle is to accomplish a particular long term goal, for example, a sporting competition or athletic event. A macrocycle can be for any time period, but is usually for a period of several months. Hayes (1995: 43) suggests that: 'Most top athletes have two macrocycles per year'.

Mesocycles

A mesocycle represents the smaller stages/blocks within the macrocycle. Each mesocycle may relate to a specific medium-term goal achievement. Generally there are four mesocycles to a macrocycle. Within sports and athletics, there are specific seasons – pre-season (conditioning and preparation), main season (competition) and post-season (recovery). Different types of training are required to meet the specific seasonal goals.

Each mesocycle has a broad training goal and aims to target specific adaptations to achieve that goal. For example, one mesocycle goal may be to chest press 50kg for 10 repetitions. The mesocycles can be planned for a different time period (see Table 9.2) but must in total add up to the timing of the macrocycle.

For example:

Table 9.1	Four example mesocycles to meet different seasonal training goals			
Mesocycle	Mesocycle	Mesocycle	Mesocycle
1–3 months	1–3 months	1–3 months	1–3 months
Medium-term goal	Medium-term goal	Medium-term goal	Medium-term goal
Conditioning	Preparation	Competition	Recovery
General preparation Training specific components – strength/endurance.	Specific preparation Training for skill, speed and psychological preparation.	Competition Where the athlete/sports person must peak and maintain form.	Recovery Where the athlete/sports person should be allowed time to recover from the competitive season.

Table 9.2	Example – varying the time period/duration of mesocycles to achieve specific goals			
Macrocycle – Long term goal – 6 months			
Mesocycle	Mesocycle	Mesocycle	Mesocycle
2 months	1 month	2 months	1 month
General conditioning	Specific conditioning	Competition and performance	Recovery
Strength and power	Speed reaction time Psychological focus	Compete	Easy training sessions
Example only! 6 weeks of strength training with weights (2 sessions per week). 3 weeks of power training combining weights (2 sessions per week) and plyometrics (1 session per week).			

Adapted from: Hayes (1995: 45) to illustrate structure rather than specific training goals or programming.

A previously sedentary individual with low levels of fitness should focus initially on developing a cardiovascular fitness foundation (mesocycle) at the start of their programme to avoid overworking. Flexibility work would need to be included as part of pre and post stretch components during this cycle. In response to the cardiovascular training, they would be likely to achieve adaptations to muscular endurance and strength, without any specific focus being given to these components. A follow up mesocycle may include specific strength and endurance conditioning work to help build the client's overall fitness and work towards their macrocycle goal.

Alternatively, a sports person having completed their conditioning and preparation mesocycles may choose one specific event in their competitive mesocycle to achieve their peak performance, such as a cup final match. Other competitive events could be used as high intensity training sessions that monitor progress towards the peak event. A recovery mesocycle would then follow the competitive mesocycle.

Microcycles

These represent how the broad training goals of each phase or stage will be achieved. They usually represent a weekly training schedule and are in most instances planned with specific training adaptations in mind. Microcycles add the detail and specifics to the training programme.

Once the microcycles are planned, the specific content for individual training sessions can be planned.

To plan a periodised programme the following will need to be established:

1 The long term goal – macrocycle
2 The medium term goals – mesocycles (seasonal goals)
3 The short term goals – microcycles, which can include the content and design of individual training sessions.

For example:

Table 9.3	**Example microcycle over 5 week period**		
Microcycles			
Week 1	Week 2 and 3	Week 4	Week 5
3 per week weights 3 day split Legs and abs Back and biceps Chest and triceps 6 reps 3 sets	3 per week weights 3 day split Legs and abs Back and biceps Chest and triceps 6 reps 3 sets	1 × weight upper body 6 reps 3 sets	Active rest

NB: session guides to illustrate structure rather than specific training programming.

Table 9.4	Example individual training sessions within a microcycle over 3 week period						
Day 1	Day 2	Day 3	Day 4	Day 5	Day 6	Day 7
50kg Deadlift	35kg Lat pull down	25kg Chest press	Rest	Repeat day 1	Repeat day 2	Repeat day 3
50kg Squats	30kg Seated row	10kg Dumbbell flyes				
20kg Lunges	20kg Bent over row	15kg Incline flyes				
15kg Leg ext	20kg Upright row	Body weight Press-ups – wide grip				
15kg Leg curl	Body weight Back extensions	Body weight Tricep dips				
Body weight Abdominal curls Oblique curls						
3 sets 10RM of each exercise	3 sets 10RM of each exercise	3 sets 10RM of each exercise		3 sets 10RM of each exercise	3 sets 10RM of each exercise	3 sets 10RM of each exercise

NB: the exercises are listed to represent how the programme can be planned rather than exact exercises or resistances used. The warm-up and cool down would also need to be planned.

Specific methods for achieving and monitoring these goals would then be planned using the specific training variables to develop different components of fitness:

- Volume – frequency, time (reps and sets) and
- Intensity – resistance, range of motion, rest, rate/speed

NB: components of fitness are discussed in part 1. Exercises and equipment that can be used within specific training sessions are discussed in part 4.

Readers wishing to develop knowledge on periodised training see Bompa and Carrera (2005) as listed in the references.

Summary

The key points discussed in this chapter include:

- Periodisation offers a method for structuring training programmes to work towards specific goals
- The three cycles within a periodised programme include: macrocycle, mesocycle and microcycle
- The duration of mesocycles can vary provided the total time equates with the duration of the macrocycle
- Individual training sessions within each micro-cycle can be planned to achieve specific short term targets/goals.

SKILLS, DRILLS AND SAFETY

This section of the book reviews some of the theories of learning that relate to the development of skill and motivation. This information is needed by the advanced circuit training instructor so that they are able to deliver safe and effective advanced circuit training sessions.

Chapter 10 reviews skill development and the theories related to learning.
Chapter 11 introduces some circuit training drills that can be used for arranging and managing the circuit training session.
Chapter 12 discusses exercise safety.

SKILL AND LEARNING

10

We constantly learn different skills throughout our lives, whether we are conscious of the process or not. Think about it – how did you learn to:

Tie your shoelaces?
Read a book?
Cook a meal?
Drive a car?
Lift a weight?

Skill is defined by Knapp (in Davis, Kimmet, Auty: 1986) as 'the learned ability to bring about predetermined results with maximum certainty, often with the minimum outlay of time and energy'.

To perform skilfully requires learning. Learning can be considered as a process where the outcome (short, medium or longer term) is a permanent change of behaviour (actions) that influence performance.

There are three main areas and domains of learning:

- Psychomotor: what an individual is able to 'do' or physically perform. Lifting a weight, kicking a ball etc
- Cognitive: what an individual 'knows'. Knowing 'how' (the technical points) to lift the weight correctly or kick the ball effectively
- Affective: what an individual 'feels' – their values and attitudes. The concept of professionalism would be developed through affective learning.

Learning a new skill

Learning a new skill requires appropriate and correct practice and repetition. Thus, with practice, performance will generally improve over time. However, incorrect practice or performance may also be learnt. Therefore to avoid or minimise incorrect learning, skills should be broken down into achievable blocks and each specific aspect practiced and learnt appropriately before moving on to the next aspect.

Another consideration would be that as new and additional skills are added, there may be a disruption to performance of other skills (short term), while new learning is taking place.

> For example:
> The technique for a bicep curl and shoulder press may be accurately performed when done as individual exercises. If combined as a sequence (co-ordination), new patterns of movement will need to be learnt and developed, which in the short term, may affect overall performance, until the new sequence is refined.

Another example would be performance of the clean. An individual may be able to perform each of the lifts that form stages towards performing the clean (deadlift, high pull, calf raise, receive) but combining the sequence will demand greater co-ordination of movement,

Table 10.1	The 'educare?' mnemonic summarises the learning needs that should be met by teachers for learners to develop skills	
	Example (from Petty: 2004)	Application to circuit training
E	Explanation of the purpose of the activity/skill and background information.	Introducing the exercise, naming the muscle working, giving key teaching points etc.
D	Demonstration – doing detail	Demonstrating correct technique
U	Use – practice of the skill	Allowing the individuals to practice the activity
C	Check and correct	Observing performance and giving feedback to correct performance
A	Aide-memoire	Providing support materials to assist learning – handouts etc.
R	Review and re-using the skill	Regular practice and reinforcement
E	Evaluation – checking the learning	Observing performance over time, asking questions regarding where the individual feels the exercise and asking if they can explain the purpose of specific exercises.
?	Queries	Giving individuals the opportunity to ask questions

The 'educare?' mnemonic relates closely to the IDEA teaching sequence introduced within the complete guide to circuit training.

Table 10.2	The IDEA teaching sequence
Introduce	Introduce the exercise – chest press Name the muscles and purpose – strengthens the muscles of the chest, front shoulder and back of the upper arm – the pectorals, anterior deltoid and triceps.
Demonstrate	Show the exercise – using a front and side demo and encourage the individuals to move around and observe from different angles.
Explain	While demonstrating, point out and explain the key technical points – the position of the spine, how to hold and position the bar, the alignment of the moving joints, the speed of the movement, the breathing pattern etc.
Activity	Let the individual try it, give reminders of the key technical points to guide perform-ance, ask where they can feel the exercise (questioning) etc. Observe and check technique and give feedback, make corrections Ask if they have any questions, offer alternatives if they struggle etc.

that also needs to be learnt to refine the movement.

Petty (2004: 23) suggests the following needs must be met when learning a new skill. He offers these needs in the form of the 'educare?' mnemonic, see Table 10.1, page 83.

Factors that may influence learning and performance

There are a number of factors that may influence learning. These include:

Psychological factors

Boredom (if an exercise is too easy), anxiety (if an exercise is too hard or perceived as too hard or if the individual feels pressured to perform in a specific way), lack of motivation (if an exercise seems unrelated to personal goals or the purpose is not explained).

Physical factors

Individual differences (body type, age, gender, levels of training, nutrition etc) will all affect performance and learning. An individual will need to be equipped with the specific physical resources to perform certain activities. For example, there would need to be some preparatory training before attempting to run a marathon. In addition, inadequate nutrition would also affect physical performance.

Mental factors

Performance of any activity demands a series of thought processes. The way an individual approaches a task mentally will influence how they perform the task and subsequent learning. Approaching a task with a positive

attitude (likelihood of success rather than failure) may not necessarily mean that the activity is performed skilfully first time, but it will affect how the individual responds to their achievement and whether they continue with subsequent attempts (try again or can't do it).

Stages of skill learning and development

Fitts and Posner (1967) suggest there are three stages to achieving skilled performance (in Reece and Walker: 2003 and in Davis, Roscoe, Roscoe, Bull: 2005).

These stages are:

- The cognitive phase
- The fixative or associative phase
- The autonomous phase

Learning theories

There a number of theories that relate to learning. A few are introduced and discussed.

Operant conditioning theory

An operant is a series of actions that a performer or learner completes (Skinner in Reece and Walker: 2003). Behaviour that occurs naturally is referred to as operant behaviour (Davis et al: 2005). To condition an operant so that it becomes a permanent behaviour requires reinforcement.

For example:
Praising (rewarding) an individual for performing an exercise correctly is likely to bring about a repeat performance.
'Well done, you kept the bar really close to your body during the deadlift.'

Table 10.3	Stages of learning to achieve skilled performance	
Phase	Level of awareness/ability	Teaching activities
Cognitive	Getting to know what is required to complete an activity. Analysis of different stages/techniques required to perform the skill (breaking down movements). May include learning the: • Purpose of activity • Technique/teaching points • Things to avoid and things to look for	Demonstration Explanations, instructions and reinforcement of teaching points to assist memorising of information. Question and answer to assist understanding. For example, why do you think it is necessary to keep the bar close to the body to perform a deadlift? Where are my knees positioned in relation to my ankles? Safety precautions
Fixative Associative	Establishing correct patterns Minimising technical errors Understanding the aims of an activity and the reasoning behind safety precautions and technique points.	Demonstrations Explanations – highlighting key points of techniques. Feedback – highlighting positive aspects and areas to develop.
Autonomous	Correct movement patterns performed automatically with skill and accuracy. Concentration can be focused on other areas – addition of other skills and/or aware of other players/athletes/game strategy.	Feedback can be specific to the performer to further refine technique. Awareness can be raised towards other factors – in sporting games, the position and actions of other players.

During the early stages of learning, more reinforcement is needed. Once a skill is learnt and permanent, reinforcement becomes less essential. The focus can be on advancing the range of skills and reinforcing the practice of these.

Positive and negative reinforcement

Positive reinforcement can generate positive feelings within the learner that may encourage them to repeat the performance. However, praise will only serve as reinforcement if it brings about positive feelings within the specific learner.

Negative reinforcement

Negative reinforcement can also bring about learning. For example, if an individual is performing the deadlift incorrectly, raising their awareness of negative effects of lifting incorrectly may assist learning.

'Be careful not to let your shoulders drop forward, this may place strain on the lower back. Can you feel the strain on your back?'

Most individuals would want to avoid a negative consequence and thus would also modify their technique. However, this will again depend on whether the learner has the internal connections and awareness to recognise how they are performing and also the extent to which the reinforcement affects them – their feeling state.

Cognitive learning theory

Cognitive learning theories suggest that individuals learn when they are able to organise information and create meaning from it – making sense of something. This involves more than just doing and receiving reinforcement. It requires reflection and thinking about what one is doing.

For example:
Skilled sports people would reflect on their game. They may review the game and their performance using a DVD. They may analyse their own and other players' movements and the consequences of these actions. The sports coach may offer feedback and input. Other players may also offer feedback and input. All of the information gathered will be used by the individual to create their own meaning.

Gestalt theory – insightful learning

Gestalt theorists suggest that learning takes place through insight – 'aha' experiences.

For example:
During a step class, trying to perform a reverse turn (or other choreographed movement) and making numerous attempts – trial and error and then suddenly finding the correct foot pattern.

Gestalt theorists also suggest that learning is most effective when the whole problem/skill is seen rather than just parts. They suggest that movement patterns are learnt more effectively when they are practiced as a whole, rather than broken down into specific stages.

For example:
When performing the clean. They would recommend the whole lift be demonstrated to show the complete picture. They would then recommend that practice occur through trial and error so that the learner can find their own solution 'aha' to aid performance.

Humanist theory

Humanistic theory suggests that individuals are constantly striving (are motivated) to be the best they can be (self actualisation). One theorist, Abraham Maslow, contributed a hierarchy of human needs, whereby the needs at lower levels need to be met before the needs at higher levels can be achieved or become the primary motivator.

Another example would be if an individual is learning a new exercise their initial need would be one of self esteem or reaching competency in that exercise, but if they became thirsty during the process, then their primary drive would change to needing to satisfy their thirst.

Another humanistic theorist, Carl Rogers, placed the learner at the centre of the process – student centred learning. He advocated that the role of the teacher was to provide appropriate

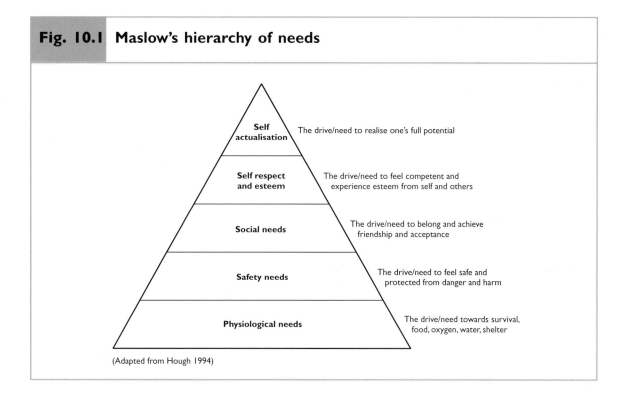

Fig. 10.1 Maslow's hierarchy of needs

Self actualisation — The drive/need to realise one's full potential

Self respect and esteem — The drive/need to feel competent and experience esteem from self and others

Social needs — The drive/need to belong and achieve friendship and acceptance

Safety needs — The drive/need to feel safe and protected from danger and harm

Physiological needs — The drive/need towards survival, food, oxygen, water, shelter

(Adapted from Hough 1994)

conditions (unconditional positive regard, empathy, and congruency) in which the individual could develop a healthy self concept and guide their own learning.

An example of inappropriate conditions would be a coaching environment where a sports person was blamed solely for losing a game, where the sports person's own feelings regarding the game were not heard or accepted and where judgements were made inappropriately relating to their person. All of these would negatively affect the individual's self-concept.

An appropriate example would be a coaching environment where all members of the team were respected and valued, successes and failures were responded to as a team contribution/effort and where each person takes responsibility for their actions while at the same time not being subjected to blame.

For example (see also fig. 10.1):
An individual would learn and perform more effectively when:

- There is water available for them during activity (physiological needs)
- The environment is safe and complies with health and safety regulations (safety needs)
- They feel part of the group and are valued and respected by their team mates and their coach (social needs)
- They feel competent to perform exercises (esteem needs)
- With lower needs met, there would be greater motivation and drive towards performing at one's best.

The role of the instructor/coach

The role of the instructor is to provide a working environment that is safe, friendly, supportive and where the group is engaged and motivated. Learning is most effective when:

- The environment is positive and supportive
- An individual wants to learn
- Individuals can contribute to their own learning and development
- There is clear understanding of the task
- The purpose for performing a task/activity are accepted by the individual(s)
- The task/skill is practiced and repeated
- Positive and meaningful (for the individual) feedback and reinforcement is provided following performance

Some specific instructing skills include:

Demonstrations

Demonstrating accurate technique to provide a role model for performance.

Social learning theory suggests that behaviour is influenced by other people. Thus, demonstration is a powerful teaching tool. (Bandura, in Davis et al: 2005). The disadvantage is that incorrect technique can also be learned. It is therefore essential that the instructor is able to provide an accurate demonstration.

Explanations

Explanations should be clear and concise. The purpose of the activity needs to be provided. In addition, it is essential to raise awareness of key technique points and factors to avoid.

Feedback

Offering positive feedback regarding performance is important.

The aim of feedback should be to promote and reinforce correct performance and raise awareness of errors in performance. The instructor should highlight:

- The purpose of performance
- Correct and incorrect techniques
- How correct movement feels
- Cause of errors
- Changes that will correct errors
- Reasons for the suggested changes
(Davis, Roscoe, Roscoe, Bull: 2005)

Questioning

The instructor will need to be able to ask questions that encourage individuals to think about their practice. Open questions (what, why, how, where?) are more effective as they promote thought and generate more than a one word answer. They also offer the opportunity to probe and ask further questions.

For example:
Where do you feel this…?
What do you think…?
How does that…?

Motivation

Another key skill of the instructor is to be able to motivate individuals. Praise and encouragement will go some of the way towards motivation. However, it is useful to know more about individuals within a group and what motivates them. Some people may need or prefer a little more 'pushing', others may need and prefer encouragement in different way. What motivates one person can be de-motivating for another.

The task of motivating individuals will be very much a learning process for the instructor and may be a trial and error process full of 'aha' experiences.

Summary

The key points discussed in this chapter include:

- A definition of skill and an introduction to the different learning domains that relate to the development of skilful practice
- Factors that influence learning and skill development
- Stages of skill learning and development
- A variety of learning theories that contribute ideas towards appropriate strategies for developing skilful practice and assisting motivation
- A list of teaching skills/strategies that a teacher can use to assist learners develop skilful performance.

ADVANCED CIRCUIT DRILLS AND DESIGN

11

There is a variety of different approaches to designing advanced circuit training programmes. The basic approaches described in *Circuit Training: A Complete Guide to Planning and Instructing* can be used (Lawrence and Hope: 2007, 2nd edition). Additionally, the following chapter provides some more advance circuit training plans that can be used. The exercises listed can be adapted to include those in part 4, which describes more advanced exercises using hardcore training equipment.

Tri sets or giant sets circuit

This circuit has three exercises for the same body region to one circuit station. Each exercise has to be completed for the desired number of repetitions or time period, prior to moving on to the next exercise at that station. No rest time is given apart from change over in between the three stations. Once all three exercises have been completed, move to the next station of the three exercises.

Table 11.1   Example tri sets circuit

Exercise	Number of reps	Exercise instructions
1. Reverse fly		1. Sit or kneeling down lean forward resting torso on thighs. Holding appropriate weight in each hand lead with elbows, keeping wrist and arms straight. Lift weight. Repeat.
2. Prone fly		2. Lying prone on either a mat or step box, arms out straight, at right angles to torso, slight bend at elbow, lift weight. Repeat.
3. Seated row weighted (wide grip)	10–15	3. Seated on mat or step box, holding a rubber band, keeping back upright and still, abdominal muscles held tight, leading with the elbows which are kept at right angles, pull hands back until level with chest.
1. Bench press		1. Lying supine on a bench, holding weight with a straight-arm above mid-chest, slight bend in elbow. Lower weight leading with elbow keeping wrist firm until weight is just above chest, extend.

Table 11.1	Example tri sets circuit (cont.)	
Exercise	Number of reps	Exercise instructions
2. Bench fly 3. Press-ups	10–15	2. Lying supine on a bench, holding weight with arms straight above mid-chest, slight bend at elbow, wrist firm. Lead the movement with the elbows, take arms out to the side. 3. Take up a prone position hands on floor wider than shoulder width apart. Maintain a neutral body position, ankles, hips, shoulder and ears. Bend at elbows and lower chest towards the floor, extend and repeat.
1. Abdominal curl 2. Russian twist 3. Lying abdominal side bends	10–15	1. Lie supine on mat, knees bent, feet flat on the floor. Place hands and arms in an appropriate position, keeping a gap between chin and chest, pull in the deep abdominal and pelvic floor muscles, lift and lower shoulders, by bending at spine. Repeat. 2. Sit on a mat in a curl-up position with knees bent and feet off the floor. Rotate the upper body from side to side (use a medicine ball). During the action maintain a neutral spine position (hips shoulder ear alignment) and tight deep abdominal and pelvic floor muscles. 3. Lying supine on a mat with arms straight resting on the floor. Lift up through the spine (semi curl-up position); knees bent feet flat on the floor. Laterally bend the spine, reach and touch right ankle, back to centre and repeat on the other side. Repeat for desired number of reps.
1. Back extension 2. Sand lizard 3. Twisting back extension	10–15	1. Lying prone on mat, hands either on bottom or under forehead. Keep hips, knees and feet on the floor. Lift the upper body and arms off the floor, head and eyes looking downwards. 2. Lying prone on a mat, arms outstretched above the head inline with shoulders. Lift alternate arm and leg alternatively, keeping abdominal and pelvic floor muscles held tight and head looking down. 3. Lying prone on a mat, forehead resting on hands. Lift upper body in the same way as in back extension, hold then twist the upper body to the right, return back to centre, lower to floor. Repeat to the left. Continue for the desired number of repetitions.

Table 11.1	Example tri sets circuit (cont.)	
Exercise	Number of reps	Exercise instructions
1. Squats		1. Stand feet hip width apart, feet facing forwards. Bending at the hips pushing bottom backwards and downwards, then bend at the knees keeping them in line with feet. Lower until thighs are parallel with the floor. Extend and repeat.
2. Lunge	10–15	2. Stand with feet hip width apart, step forward with the leading leg, keeping knee in line with feet. When foot hits the floor bend the back knee down until both knees are at a right angle. Push off the front foot and regain hip width stance, repeat with the other foot. Continue for the desired number of repetitions.
3. Step-ups		3. Using a step box or weights bench (18inchs) stand behind bench and place the leading foot wholly on the step, push off the rear leg and step onto the step, achieve full extension through the body before stepping off, repeat using alternate leading leg.
1. Calf raise		1. Stand feet hip width apart. Full natural extension through the whole body. Keeping a slight bend in the knees, lift heels off the floor as high as possible, lower. Repeat for desired number of repetitions.
2. Single leg – right	10–15	2. Stand on one foot rest the other behind, use a wall for support and balance. Lift up through the foot as above.
3. Single leg – left		3. As above
1. Triceps ext		1. Sit on a step box or weights bench holding a dumbbell/rubber with two hands above the head on straight arms. Bend at the elbow and lower resistance behind the head, keeping the elbows pointing upwards. Extend keeping a slight bend in elbow at top of movement, repeat for desired number of repetitions (maintain upright posture abdominal muscles held tight).

Table 11.1	Example tri sets circuit (cont.)	
Exercise	Number of reps	Exercise instructions
2. Triceps dips	10–15	2. Begin sitting on a step or weights bench with hands next to thighs, fingers facing forwards. Balance on your arms, moving backside in front of the step with legs straight (harder) or bent (easier). Bend the elbows and lower body, keeping the shoulders away from your ears and the elbows parallel to one another, going no lower than 90 degrees. Push back up to starting position, repeat for desired number of repetitions.
3. Triceps pushdown		3. Using two rubber bands fix one end over the top of a door and close it or to a fixed point on the wall above the head. Stand facing the door/wall feet hip width, body upright with tight abdominal and pelvic floor muscles. Keep upper arm locked into the side. Holding onto the rubber bands, extend the lower arm until straight (slight bend in elbow) return under control to start position, repeat for desired number of repetitions.
1. Lateral raise	10–15	1. Stand upright, feet wider than hip width. Using either a rubber band or dumbbells, held down by the side on a straight arm (slight bend in elbow) lift from the shoulder out to the side until arm is level with the shoulder (elbow high). Lower under control to the start position. Repeat for desired number of repetitions.
2. Frontal raise		2. Stand upright feet wider than hip width. Using either a rubber band or dumbbells, held down by the side on a straight arm (slight bend in elbow), raise the resistance in line with the shoulder to the front until level with shoulder. Lower under control, repeat for the desired number of repetitions.
3. Shoulder press (reduced range of movement)		3. Stand or sit with an upright posture, hips in line with shoulders and ears, abdominal and pelvic floor muscles held tight. Hold the resistance at shoulder level elbows pointing to the floor, knuckles to ceiling. Punch the knuckles upwards until the upper arm is perpendicular to the floor, lower to start position under control, repeat for desired number of repetitions.

Table 11.1	Example tri sets circuit (cont.)	
Exercise	Number of reps	Exercise instructions
1. Modified straight leg deadlift		1. Stand feet hip width apart with a small bend in the knee, holding a suitable weighted barbell. Push bottom backwards, lean forwards from the hips until bar crosses the knees, keeping arms and back straight. Abdominal and pelvic floor muscles must be pulled in tight. Return to the upright position; repeat for the desired number of repetitions (knees remain bent throughout).
2. Hip extension		2. Lie supine with shoulders resting on a mat with feet hip width apart resting on a high step or stability ball, arms on the floor palms down for support. Push up through the hips until the body is straight. Lower to the starting position and repeat for the desired number of repetitions.
3. Leg curl		3. Lie prone on a bench or a mat. Place a suitable weighted dumbbell between the feet or a rubber band round the ankles with the other end fixed to a point or behind a closed door. Bend at the knees drawing the heels towards the bottom. Return under control, repeat for the desired number of repetitions.

Compound circuit

This circuit works more than one area of the body by allowing the exerciser to perform two exercises in one. It can be a very challenging circuit, providing each exercise is performed correctly with an appropriate weight or resistance challenge. Use a slow to moderate speed for all exercises. Perform more than one circuit to add further challenge.

The pack of cards circuit

This circuit involves using a pack of playing cards. Give the cards a good shuffle and place them face down on the floor. Each card is taken off the top and turned over one at a time. Whichever suit is shown (hearts, clubs, diamonds or spades) will represent, an exercise:

For example:
Hearts   = Press-ups
Clubs   = Squats
Diamonds   = Curl-ups
Spades   = Back extension
Jokers   = 8 count Burpee* for
              10 repetitions

*Stand upright with feet hip-width apart (1), then squat, placing hand on the floor under shoulders (2). Shoot legs back until in the press-up position (3). Perform one complete press-up (4). Straddle legs out to side (5). Then draw back to press-up position (6). Jump back to squat position, knees between arms (7). Stand up (8).

Perform the pack of cards exercise for the number of repetitions indicated by the specific card value/number:

Aces count as 11 (same as jacks)
Jokers can have their own designated exercise!
Picture card value is optional
Jacks – 11
Queens – 12
Kings – 13
(NB: value of picture cards can be adapted at discretion of instructor, but should be above 10)

Continue through the pack until all 54 cards, including the jokers, have been turned, and exercises performed. Exercises can be changed at any time to maintain progression and interest.

Progression/modification/alternative

Start by using card values one to six. With progression, higher value cards for every suit can be added every one or two weeks until the entire pack is being completed.

An alternative way of performing this circuit is to have only two exercises, one for red cards and one for black. Again repetitions are performed that are equal to the card value.

Table 11.2	Example compound circuit	
Exercise	Reps/time	Exercise instructions
1. Lunge with biceps curl	10–15 reps 35–45 seconds	Standing with feet hip width apart, holding a set of dumbbells. Step forward and lunge while performing a biceps curl.
2. Triceps dip with hip flexion	10–15 reps 35–45 seconds	Rest hands on a bench, fingers facing forwards, bottom just off the bench, legs bent or straight. Bend at elbow, lower body until upper arm is parallel to the floor, at the same time bend at the hips and lift one leg in the air. Return to start position, repeat using alternate legs.
3. Back extension with prone fly	10–15 reps 35–45 seconds	Lie prone on a mat, upper arm at right angles to torso and elbows bent at right angles. Lift upper body off the floor at the same time pull back on elbows, return to start position, repeat.
4. Abdominal curl with leg curl	10–15 reps 35–45 seconds	Lie supine on a mat with legs held straight together up in the air, hands on temples. Lift shoulders off the floor, at the same time bend at the knees and lower heels towards the bottom return to starting position, repeat.
5. Squats with calf raise	10–15 reps 35–45 seconds	Stand with feet hip width apart, bend at hips and knees until bottom is level with knees, extend through knees and hips back to a standing position, continue upwards by lifting heels off the floor, return to starting position, repeat.

Table 11.2	Example compound circuit (cont.)	
Exercise	**Reps/Time**	**Exercise instructions**
6. Press-ups with hip extension	10–15 reps 35–45 seconds	Take up a front support position, hands wider than shoulders. Bend at the elbows and lower body until chest is near touching floor at the same time extend through the hips with one leg, straighten the arms, lower the leg, repeat using alternate legs.
7. Seated row with triceps extension	10–15 reps 35–45 seconds	Sit upright on a mat legs straight in front. Wrap a rubber band around the feet taking hold of each end. Leading with the elbows, keeping arms bent and close to the body, extend until full extension of the shoulder then extend through the elbow, return to start position, repeat.
8. Toe taps with wrist curls	10–15 reps 35–45 seconds	Stand feet hip width apart, body upright. With straight arms out in front of you, hold a foot long bar with string and a weight tied to the weight and bar. Start lifting each foot (dorsi flexion) off the floor alternatively, while winding the string on and off the bar.
9. Hip extension with chest fly seconds	10–15 reps 35–45	Lie supine on a mat, rest feet on a bench or stability ball, push hips forwards until in line with shoulders and knees. At the same time perform chest fly with a suitable size dumbbell. Return to start position, repeat.
10. Multi-abdominal	10–15 reps 35–45 second	Lie supine on a mat; knees bent feet flat on the floor, arms straight palms flat on the floor. Curl up through the spine and hold this position, laterally flex to the right to touch the right ankle, back to centre then to left ankle. Reach across to touch the outside of the right knee, back to centre then touch the outside of the left knee, back to centre, lower to floor, repeat.
11. Pinch grip farmers walk	Set distance walked	Holding on to suitable weight discs by pinching between thumb and fingers in each hand, walk with arms straight down by the side of the body for the set distance or time.

Add on circuit

Have all exercise cards and equipment ready, laid out in a suitable training area.

- The idea of this circuit is to turn over one card, perform the exercise listed and then walk or jog a short distance as a transition period before returning to turn over another card
- Perform the first exercise again and add on

the second exercise, repeat the transition exercise (walk or jog) again
- Then perform the first and second exercises, adding on the third exercise, determined by turning over a third card
- Complete this format until all exercises have been completed

10 reps Triceps dip
10 reps Upright rows
10 reps Deadlift
10 reps Bench fly
10 reps Step-ups
10 reps Skaters
10 reps Abdominal curl
10 reps Squat thrust
10 reps Calf raise 5x5
10 reps Reverse fly

Complete this circuit using correct form and you are worthy!

Progression/modification/ alternative

- Start the circuit with fewer exercises adding extra exercises at each session
- Increase or decrease the repetitions to be completed up to a max of 13
- Adjust the intensity of each individual exercise (easier/harder)
- Change the order of exercises
- Increase or decrease the length of the transition distance/time

The ton-up circuit (100 reps circuit)

This particular circuit is a great favourite among members of the armed forces. It requires the exerciser to perform 100 repetitions of any single exercise at every station during the

Ideas

Question:
Want to increase your natural body resistance to give you a better workout? Want to weigh more than you do, without having to buy expensive weight vests etc?

Answer:
The humble rucksack may be the answer.
Load the rucksack up with magazines, which are easy to mould, place rucksack tightly on your back, better if you have a chest strap that can be tightened up and away you go. If magazines are not heavy enough use phonebooks, or a combination of both. Try using this idea when performing some of the exercises in this book.

Question:
Want to know what to do with a stability ball once it has passed its sell by date when it's not suitable to sit or lie on any more?

Answer:
Recycle it!
Fill the stability ball half or three quarters full of water, place the stopper back in and you now have a water ball – great for lifting and carrying. This will definitely help improve core stability and the bracing of the deep abdominal muscles. You could even substitute the sandbag (see part 4, chapter 15) with one of these for some of the exercises shown in this book.

session. Repetitions can be accumulated over the session or done all at once depending on the individual's fitness level, but the end goal is to achieve 100 repetitions.

Initially, select compound large muscle group exercises such as deadlift, squat, lunge, press-ups, bench press, bent forward rowing, or seated rowing. Whether using resistance or not, ensure that a whole body approach is achieved. Over time add to the list, but remember the more exercises are in the programme the longer the session is likely to be. One hundred repetitions of each exercise is the final goal. Fewer repetitions, depending on fitness level, could be used to start out with, say around 50, building up to the targeted 100.

Alternative ton-up circuit (100 reps) approach

An alternative approach is to perform as many repetitions of the exercise as you can, using correct technique, without stopping. When you have to stop, because you can no longer maintain correct technique, note the number of repetitions you have completed (35 reps) take this number away from 100 and that is the amount of rest you are entitled to (100 − 35 reps = 65 sec rest). After the rest continue to perform the same exercise again until you stop (i.e. achieved 20 reps). Add the first number of repetitions to the second (35 reps + 20 reps = 55 reps). Again take this number away from 100 and that's now your rest interval (45 sec rest). Continue this format until you have completed/accumulated 100 repetitions of the exercise, then move to the next exercise. Again start on large compound exercises ensuring a whole body balanced approach in the workout session.

A 22-minute workout circuit approach

This circuit workout requires the exerciser to complete 8 sets of 8 exercises in a target time frame of 22 minutes. Equipment required is a weights bar and plates, pencil and paper to record sets completed and a clock.

Make sure that a suitable warm-up is completed before the start of the 22-minute workout.

Select a suitable weights bar, which will be used for the whole circuit (12kg–15kg is suitable, a lighter weight should be used initially until the routine has been learnt). Complete the following exercises without releasing your grip on the bar:

- Shoulder press
- Squat
- Biceps curl
- Deadlift
- Upright row
- Burpee
- Bent over row
- Press-up

Each exercise is performed for 10 reps, with no rest inbetween. After the completion of one complete set place the bar on the floor and record the set completed. (This is the only rest that is allowed; the time it takes to put the weight down and mark the set number completed and pick up weight again.) Carry on with the next set and repeat the workout as before until you have completed a total of 8 sets. Take a note of the time it has taken to complete all 8 sets, when the final exercise of the final set has been completed. Target time is 22 minutes. If this time has been beaten then increase the weight the next time.

300 repetitions circuit approach

The current trend for doing extreme repetition circuits comes on the back of the Hollywood blockbuster *300*. The film is about how 300 Spartan warriors lead by King Leonidas fought and died at the battle of Thermopylae, against Xerxes and his massive Persian army in 480 B.C. For authenticity the actors had to be shaped as Spartan warriors which meant extreme strength and conditioning training. The 300 repetitions workout was the brain child of Mark Twight, a self taught exercise guru and former world-class mountain climber, who apparently still clings to the 'no pain, no gain' mantra. The actors and stuntmen were trained at Gym Jones, his invitation-only, no-frills gym in downtown Salt Lake City, where 'there's no air conditioning, no mirrors, and no place comfortable to sit' (K. Doheny, 2007). So what was the workout? The following is the end goal (the end of the macrocycle if you like):

- 25 pull-ups
- 50 reps deadlifts 61kg
- 50 reps push-ups
- 50 reps box jumps 61cm box
- 50 reps floor wipers – a core and shoulders exercise at 61kg
- 50 reps clean & press 16kg
- 25 pull-ups
  Total of 300 reps

Each exercise has to be performed with 100% correct technique and full range of movement with no rest between exercises. The final score is based on total time taken. Regular training, over a period of six months using extreme circuit training activities as described in this book, can prepare you to complete this workout. Training

for the actors required 90 minutes to two hours a day, five days a week. Stuntmen trained 90 minutes to two hours, five days a week. The workout should be completed in less than 20 minutes.

Alternative 300 circuit approach

The following circuit is more achievable than the previous one but still relies on the exerciser completing a total of 300 repetitions in any single set. This approach has a total of 10 exercises.

- 30 reps jump squat
- 30 reps staggered push-ups
- 30 reps alternate spit jumps
- 30 reps plyo press-ups
- 30 reps 'prison squats'
- 30 reps hindu press–ups
- 30 reps kneeling band row
- 30 reps Burpees
- 30 reps pull-ups
- 30 reps V-sits

Each exercise has to be performed using correct technique and full range of movement, with no rest between exercises. Take note of the time it takes to complete the whole circuit. Try to better it over time. Exercises and their intensity can be changed to suit individuals. Progress to the 300 via 100 then 200 repetition circuit using the same format as above.

Summary

The circuit designs described in this section include:

- Compound circuit
- Pack of cards circuit
- Ton up circuit
- Alternative ton up circuit
- 22 minute circuit
- 300 rep circuit
- Alternative 300 rep circuit.

SAFETY CONSIDERATIONS

12

Many of the exercises illustrated and explained in this book are of an advanced nature. They would not be appropriate for beginners to exercise or people who are unused to exercise and/or who have specialist considerations (medical conditions etc).

Instructors must use their professional judgement to assess the suitability of any exercise for specific individuals. This would include gathering information regarding the individual's exercise history and current fitness level (discussed in the *Circuit Training: A Complete Guide to Planning and Instructing*). As a rule 'if in doubt, leave it out'.

There is always an element of risk attached to moving the body (strains, sprains, fainting, aggravating a medical condition, heart attack). There are equally a number of risks attached to not moving the body (CHD, obesity, high blood pressure, contributing to the onset of certain medical conditions, heart attack etc).

The following offers a method for reviewing and analysing the safety and effectiveness of specific exercises and reviewing their appropriateness for different individuals.

- Speed of movement/exercise
- Environment
- Equipment (and clothing)
- Stability of exercise position
- Alignment of the joint and posture
- Weight (body weight and resistance)

If the risks attached to performing an exercise are higher than the benefits, the exercise should probably be removed from the programme. Individual factors will also need to be considered. For a start, an appropriate and beneficial exercise for a trained athlete or sports person may be totally unsuitable and carry high risk for an untrained person. Other individual factors that will affect safety and appropriateness of exercises include fitness level, skill level, age, gender, height, weight, lever length, body type, body composition, medical conditions, state of mind, reason for participating etc.

Exercises need to be modified to accommodate individual differences. This can be achieved by manipulating any of the variables that affect exercise intensity – repetitions, resistance, rate, range of motion etc.

Speed

When exercises are performed quickly, momentum is generated. Exercises performed with excessive speed may be higher risk because there is the potential for the joints to move beyond their normal range of motion and without sufficient control. This, potentially, makes the tissues around the joint (ligaments and tendons) more vulnerable.

Individuals with longer levers and those with larger and heavier body frames will generally need more time to achieve a full range of movement than individuals with shorter levers and individuals with smaller and lighter body frames. The speed/rate of exercise will need to be modified to accommodate such differences.

Ideally, individuals should be encouraged to work at a pace that feels comfortable for them to perform. People training specifically to develop speed, can achieve this over time. The initial focus should be on working with correct technique. Once this is developed the pace can be progressed.

Environment and equipment

There are numerous factors to consider in relation to safety within the circuit training environment and the equipment used. These include:

- Space around stations
- Positioning of stations
- Positioning of equipment
- Storage and maintenance of equipment (before, during and after use)
- Lifting and handling of equipment (before, during and after use)
- Temperature
- Floor surface
- Ceiling height
- Air conditioning etc.

Some basic considerations are:

- High impact jumping movements should be performed on a sprung floor surface to reduce stress on the joints
- Correct footwear should be worn to assist with shock absorption. Shoe laces should be tied
- Equipment should be lifted and moved safely, using deadlift techniques
- There should be adequate space between stations

- The number of participants should be appropriate to the size of the training area (depending on whether the exercise is static or if there is considerable movement), the layout of stations and to the equipment used
- Equipment needs to be securely placed by users after they have performed each station
- Additional assistant instructors should be employed when working with larger group numbers or using larger outdoor spaces
- Equipment must be regularly checked for wear and tear and replaced when appropriate

Stability

All exercises should be performed using a stable base. Freestanding exercises using resistance should be performed with the feet hip width and half apart and the knees unlocked to assist balance. A narrow foot stance with locked knees would reduce stability and balance.

When performing bench exercises (chest press), some individuals raise their feet onto the bench to prevent hollowing of the lower back. Placing the feet on the bench will make the exercise less stable. It is advisable to use a step with risers at the end of the bench, to correct the alignment of the spine and offer greater stability.

When performing exercises that use a narrow base (quadriceps stretch on one leg), the support of a prop, such as a wall or partner, will assist balance until core stability is improved.

Alignment

The joints should always be moved through their appropriate range of movement to maintain the integrity of the joint structure. Movements that take the joint out of

Fig. 12.1

Chest press
(a) Feet on bench – less stable
(b) Feet supported on riser for stability – wider base of support

alignment will place the ligaments under stress and increase the risk of injury.

Movements that encourage the knees to roll inwards and/or move excessively forward of the foot (squats and lunges) can contribute to stress in the knee joint. It is essential that correct teaching advice is given to promote correct alignment.

Correct alignment of the spine should also be encouraged. The spine should be lengthened, with the head aligned, the shoulders relaxed and down away from the ears, the pelvis positioned neutrally so the natural curve of the lumbar spine is present and the abdominals should be engaged to support alignment.

Weight

Adding weight to any movement (longer levers, dumbbells and barbells, sandbags etc) will automatically increase the potential stress on the body. Exercises using external resistance should be performed with control. In addition, the resistance and weight lifted should be progressively and steadily increased so that the tissues surrounding the joints (muscles, tendons, ligaments) are able to adapt to training demands.

To minimise the risk of repetitive strain, the exercises and activities should be varied. During a session a variety of movements should be included. High impact activities such as running and jumping increase the gravitational forces through the body, increasing the stress on joints. They should be combined with lower impact activities (walking, squatting) to reduce this stress. Individuals who are heavily overweight, less fit or have an injury should not perform high impact work, as their joints would be more at risk.

Within a weekly training programme, specific activities should be varied to prevent over-use injuries. For example, cross training (walk, swim, cycle). There should also be adequate rest between training sessions for the body to recover.

Summary

The key points relating to exercise safety discussed in this chapter include:

- Speed of exercise
- Environmental considerations
- Equipment considerations
- Stability of exercise position and joint
- Alignment of joints and spine
- Weight – how the addition of external resistance affects safety.

ADVANCED HARDCORE CIRCUITS

The aim of this section is to introduce a range of equipment and exercises that can be included as part of an advanced circuit training exercise repertoire. The exercises described are designed for people who are trained and familiar with traditional exercise regimes. They are not appropriate for beginners without significant modification.

Modification can be achieved by considering the variables discussed in part one (repetitions, rate, range of motion and resistance etc).

Ideas for modification, progression and adaptation are suggested for all exercises but are not exhaustive.

- Chapter 13 Kettlebell training with exercises.
- Chapter 14 Medicine balls with exercises.
- Chapter 15 Sandbag training with exercises (power bags and X bags).
- Chapter 16 Sledgehammer training with exercises.
- Chapter 17 Rubber tyre training with exercises.
- Chapter 18 Rope climbing training with exercises.
- Chapter 19 Telegraph pole training with exercises.

KETTLEBELL TRAINING WITH EXERCISES

13

Kettlebells are small cast iron weights that resemble an old fashioned kettle. They have a flat under surface, a round body and an arched handle. Advocates of kettlebell training describe them as cannonballs with handles (Siff: 2003a, b, Tsatsouline: 2002, Yessis & Trubo: 1987).

Kettlebells come in a range of resistances to accommodate different levels of fitness and meet different training needs and requirements.

Kettlebell training originates in Russia, where they have been used for over 100 years as a training device with Olympic athletes, martial artists and the military forces. Over recent years, they have attracted the attention of the UK fitness market. Although previously used predominantly by men, they are now becoming a popular training device with women. A number of movie stars are reported in various national magazines to be using kettlebells as part of their fitness regimes.

Kettlebell exercises

| Exercise 13.1 | Double leg deadlift |

Start position

End position

NB: this exercise should be used whenever lifting kettlebells from the floor. It allows the weight to be carried by the stronger thigh muscles thereby preventing injury to the back that can occur when lifting incorrectly.

Purpose

To work the buttock muscles (gluteus maximus), the back muscles (erector spinae) and the muscles at the front of the thigh (quadriceps)

Starting position and instructions

- Position the feet at hip width and place the kettlebells outside the hips
- Bend at the knees and hips and take hold of the kettlebells
- Keep the trunk fixed
- Lift the kettlebells from the floor by straightening the knees and hips and leading the movement with the shoulders

Coaching points

- Ensure that the back is straight and abdominals pulled in
- Take care not to hollow the back
- Push the buttocks backward when lowering and avoid letting the knees travel too far forward
- The backside should be higher than the knees when bending and the shoulders higher than the buttocks
- Look forwards and ahead
- Ensure the body moves to a fully extended position without locking the joints

Progressions/adaptations/variations

- Progress to a heavier weight
- Increase repetitions to challenge endurance
- Vary the speed
- Combine with a calf raise to add variety to the movement
- Combine with an upright row to add variety to the movement

Exercise 13.2	Single leg deadlift

Purpose

To work the buttock muscles (gluteus maximus), the back muscles (erector spinae) and the muscles at the front of the thigh (quadriceps)

Starting position and instructions

- Deadlift the kettlebells to the thighs
- Hold the kettlebells, keeping the arms straight
- Keep the trunk fixed
- Extend one leg behind the body, keeping the hips square
- Bend to front leg at the hip and knee (one legged squat action)
- Letting the rear leg extend back further

Coaching points

- Ensure that the back is straight
- Keep the abdominals pulled in
- Keep the shoulders and hips square
- Look forwards and ahead
- Lower as far as is comfortable and where balance can be maintained
- Ensure the body moves to a fully extended position without locking the joints

Progressions/adaptations/variations

- Progress to a heavier weight
- Increase repetitions to challenge endurance

| **Exercise 13.3** | **Kettlebell good morning with bent knees** |

Purpose

To work the buttock muscles (gluteus maximus), the back muscles (erector spinae) and the muscles at the back of the thigh (hamstrings)

Starting position and instructions

- Deadlift one or both kettlebells to the thighs
- Hold the kettlebell(s), keeping the arms straight
- Keep the trunk and shoulder girdle fixed
- Bend the knees slightly and tighten the abdominals
- Bend forward from the hips
- Extend upwards keeping the knees slightly bent
- Repeat for the desired repetitions

Coaching points

- Ensure the pelvis is neutral
- Keep the abdominals pulled in
- Keep the shoulders and hips square
- Look forwards and ahead
- Make sure the back does not arch, visualise the scapula sliding towards the buttocks
- Move with control

Progressions/adaptations/variations

- One kettlebell can be held with both arms, between the legs
- Two kettlebells can be held at shoulder width
- Progress to a heavier weight(s)
- Increase repetitions to challenge endurance
- Raise the feet on steps to increase range of motion. Be aware of additional stress placed on the muscles

Exercise 13.4 — Upright row with kettlebell

Start position   End position

Purpose

To work the muscles at the front of the shoulder (anterior deltoid), the front of the upper arm (biceps and brachialis) and the top of the back (trapezius)

Starting position and instructions

- Deadlift the kettlebell to the thighs using an over hand grip
- Widen the foot stance to hip width and a half apart
- Hold the kettlebell in front of the body, palms face towards the body
- Keep a firm grip of the kettlebell with the thumbs tucked under and wrist fixed
- Raise the kettlebell to chest level, keeping it close to the body
- Lower under control
- Repeat for the desired number of repetitions

Coaching points

- Keep the kettlebell close to the body
- Keep the abdominals pulled in tight
- Keep the movement controlled
- Tuck the bottom under and keep the knees unlocked
- Lead the movement with the elbows, lift them as high as is comfortable

Progressions/adaptations/variations

- Progress to a heavier weight
- Increase repetitions to challenge endurance
- Vary the speed

Exercise 13.5 — Calf raise with kettlebells

Purpose

To work the muscles at the back of the lower leg (gastrocnemius and soleus).

Starting position and instructions

- Start with the feet positioned at hip width apart
- Deadlift the kettlebells to the thighs using an over hand grip
- Rise onto the ball of the foot, lifting the heel from the floor
- Lower down under control
- Perform for the desired number of repetitions

Coaching points

- Keep the kettlebells close to the side of the body
- Keep the abdominals pulled in tight and the back straight
- Keep the knee joint unlocked
- Press onto the ball of the foot with the weight central, take care not to roll the ankles outwards
- Keep the movement smooth and controlled

Progressions/adaptations/variations

- Progress to a heavier weight
- Increase repetitions to challenge endurance
- Vary the speed
- Combine with a calf raise to add variety to the movement and as a method for introducing the clean

| Exercise 13.6 | Kettlebell lunge |

Purpose

To work the buttock muscles (gluteus maximus) and the muscles at the front of the thighs (quadriceps)

Starting position and instructions

- Deadlift the kettlebells to the thighs
- Check that the feet are positioned comfortably at hip width apart, toes facing forward
- Take a large step forward and bend the knee to lower the body weight downwards
- Ensure both knees are positioned at right angles (90 degrees)
- Push through the thigh to lift the body back to an extended position
- Perform either by alternating legs or repeating the movement on the same leg
- Deadlift the kettlebells to floor on completion of the exercise

Coaching points

- Ensure the front knee does not over shoot the toe
- Step forward and sink the body down centrally rather than diving forward
- Keep the abdominals pulled in tight and the chest lifted
- Ensure the knee does not roll inwards
- Bend the back knee towards the floor but ensure the knee does not hit the floor
- Look straight ahead
- Drive through the thigh to return the body to an upright position
- Keep a relaxed grip on the kettlebells and keep the shoulders relaxed and pressed down

Progressions/adaptations/variations

- Perform through a smaller range of motion initially
- Start the exercise at a slower pace so that the muscles have to contract and work for longer
- Start with lighter kettlebells and progressively increase resistance
- Increase repetitions
- Lunge with kettlebells at side of body
- Lunge with kettlebells resting in shoulder groove
- Overhead press and then lunge
- Vary the speed of the exercise two slow double time and four normal pace
- Alternating legs will be slightly easier than repeating a number of repetitions on the same leg which will require greater muscular endurance

Exercise 13.7	Kettlebell curls

Purpose

To work the muscles at the front of the upper arm (biceps and brachialis)

Starting position and instructions

- Deadlift the kettlebells to the thighs
- Widen the foot stance to shoulder width and a half apart and unlock the knees
- Fix the elbows into the side of the body
- Keep the body lifted and buttocks tucked under
- Curl the kettlebells in an arc-like motion towards the chest
- Lower to thighs under control and fully extend (not locking) the elbow

Coaching points

- Raise the kettlebells under control
- Keep the abdominals pulled in tight
- Keep the wrist fixed and straight

- Keep the elbows and upper arms close to the body
- Avoid locking the elbow

Progressions/adaptations/variations

- Increase repetitions or resistance
- Perform at a slower pace so that the muscles have to contract and work for longer
- Vary the speed of the exercise two slow double time and four normal pace
- Perform through different ranges of motion
- Concentration curls either seated or squatting

Exercise 13.8	Kettlebell clean

Start position — End position

and soleus), the shoulder muscles (deltoid), the middle and upper back muscles (trapezius), and the muscles at the front of the upper arm (biceps). To position the kettlebells at chest level for overhead pressing exercises

To challenge cardiovascular fitness when performed sequentially

Starting position and instructions

- Deadlift the kettlebells to thighs
- Receive: squat and swing the kettlebell(s) to groove of shoulder
- Return: swing press the kettlebells away from shoulders and bend the knees into squat position
- Repeat for desired number of repetitions
- Reverse deadlift to lower to floor

Coaching points

- Keep the back straight
- Keep the abdominals pulled in throughout the movement
- Keep joints unlocked

Progressions/adaptations/variations

- Perform the exercise without kettlebells to get used to the movement
- Aim to perform as a smooth and fluid movement
- Single arm clean
- Alternating clean
- Double arm clean
- Clean and press (variations as above)

Purpose

To work many of the major muscles. The buttock muscles (gluteus maximus), the back muscles (erector spinae), the muscles at the front of the thigh (quadriceps), the calf muscles (gastrocnemius

| Exercise 13.9 | Kettlebell wide legged squat |

Purpose

To work the buttock muscles (gluteus maximus) and the muscles at the front of the thigh (quadriceps)

Starting position and instructions

- Clean the kettlebells to rest in the groove of the shoulder
- Position the feet shoulder width and half apart
- Bend the knees to squat and lower the body downwards
- Return to an upright position to complete the lift
- Perform for the desired number of repetitions
- To return the kettlebells to floor, reverse the clean and deadlift actions

Coaching points

- Look forward throughout the movement
- Keep the knees travelling in line with the feet
- Do not let the bottom drop below the knees when squatting downwards
- Keep the back straight
- When straightening the knees do not lock the knees
- Fully extend the hip and knees

Progressions/adaptations/variations

- Increase resistance and/or repetitions
- Perform at a slower pace so that the muscles have to contract and work for longer
- Perform through a smaller range of motion initially

| Exercise 13.10 | Kettlebell narrow legged squat |

Purpose

To work the buttock muscles (gluteus maximus) and the muscles at the front of the thigh (quadriceps)

Starting position and instructions

- Deadlift one kettlebell and hold slightly in front of the body with elbows bent
- Position the feet hip width apart
- Bend the knees to squat and lower the body downwards, pushing the bottom back
- Return to an upright position to complete the lift
- Perform for the desired number of repetitions
- To return the kettlebells to floor, reverse the clean and deadlift actions

Coaching points

- Look forward throughout the movement
- Keep the knees travelling in line with the feet
- Do not let the bottom drop below the knees when squatting downwards
- Keep the shoulders lifted
- Maintain neutral pelvis and abdominals pulled in
- When straightening the knees do not lock the knees
- Fully extend the hip and knees

Progressions/adaptations/variations

- Increase resistance and/or repetitions
- Perform at a slower pace so that the muscles have to contract and work for longer
- Perform through a smaller range of motion initially

Exercise 13.11 — **Kettlebell swing**

Start position | End position

Purpose

To work the buttock muscles (gluteus maximus) and the muscles at the front of the thigh (quadriceps) and the front shoulder (anterior deltoid)

Starting position and instructions

- Deadlift one kettlebell
- Widen the stance to squat position and hold a kettlebell between the knees, bending at hips and knees
- Keeping the arms extended, the elbows unlocked
- Swing the kettlebell forward and upwards to shoulder height, fully extending hips and knees
- Lower by swinging back through the legs
- Perform for the desired number of repetitions

Coaching points

- Maintain control throughout
- To return the kettlebells, use deadlift

- Look forward throughout the movement
- Do not let the bottom drop below the knees when squatting downwards
- Keep the back straight, abdominals tight and shoulders relaxed
- When straightening the knees do not lock the knees
- Fully extend the hip and knees

Progressions/adaptations/variations

- Increase resistance and/or repetitions
- Perform through a smaller range of motion initially

Exercise 13.12	Kettlebell shoulder press

Start position   End position

Purpose

To work the shoulder muscles (deltoids), the muscles of the upper back (trapezius) and the muscles at the back of the upper arms (triceps)

Starting position and instructions

- Clean the kettlebells to the rest into the groove of shoulders
- Press the kettlebell over head
- Lower the kettlebell under control
- Repeat for the desired number of repetitions

Coaching points

- Keep the back straight and abdominals pulled in
- Keep the knuckles facing up and the wrist fixed
- Take care not to hollow the back or lock the elbows
- Keep the knees slightly bent throughout the movement
- Keep a smooth and comfortable movement

Progressions/adaptations/variations

- Increase repetitions or resistance
- Alternating press
- Single arm press
- Double arm press
- Squat and press
- Press and lunge (press kettlebell overhead and lunge)
- Perform on a stability board (very advanced)

Exercise 13.13 | **Kettlebell lying chest press**

Purpose

To work the muscles at the front of the chest (pectorals), the muscles at the front of the shoulder (anterior deltoids) and the muscles at the back of the upper arm (triceps)

Starting position and instructions

- Lie on the floor with either knees bent or legs straight
- Neutralise the pelvis
- Hold the kettlebells into the grooves of the shoulder
- Push the kettlebell straight up to the ceiling
- Lower under control
- Perform the desired number of repetitions

Coaching points

- Keep the knuckles facing upwards and the wrist fixed
- Keep the abdominals pulled in tight and take care not to hollow the back
- Move the kettlebells in a straight line, level with the chest
- Extend the arms fully but do not lock the elbows
- Keep the elbows and wrists in line and vertical
- Keep the movement smooth and under control

Progressions/adaptations/variations

- Increase repetitions and/or resistance
- Vary speed
- Single arm press
- Alternating press
- Double arm press

Exercise 13.14 | Kettlebell bent over row

Purpose

To work the muscles at the back (latissimus dorsi) and the muscle at the front of the upper arm (biceps and brachialis)

Starting position and instructions

- Wide or narrow squat position
- Deadlift kettlebells
- Bend forward at hips and bend knees, extending spine
- Row the kettlebell(s) to the armpit keeping it close to the body
- Lower under control, extending arm
- Repeat for the desired number of repetitions

Coaching points

- Keep the back straight and look forward and slightly down throughout
- Keep the abdominals pulled in tight
- Ensure the elbow does not lock as the kettlebell lowers
- Take care not to twist the back
- Keep the shoulders square

Progressions/adaptations/variations

- Increase repetitions and/or resistance
- Vary speed
- Single arm row – one side at a time
- Alternating row
- Double arm row

Exercise 13.15	Kettlebell renegade row

Coaching points

- Keep the spine neutral
- Keep the abdominals pulled in tight
- Ensure the elbow does not lock as the kettlebell lowers
- Keep the kettlebell close to the body
- Take care not to twist the back
- Keep the shoulders square

Progressions/adaptations/variations

- Increase repetitions
- Increase resistance
- progress from $^3/_4$ to full press-up start position
- Raise the feet or knees onto a bench

Exercise 13.16	Sit-ups/curl-ups with kettlebells

Purpose

To work the muscles at the back (latissimus dorsi) and the muscle at the front of the upper arm (biceps and brachialis)

Starting position and instructions

- Adopt a press-up position ($^3/_4$ or full) with hands holding the handle of kettlebells
- Lift one kettlebell off the floor using a rowing action
- Lower under control to the floor
- Repeat with other arm
- Repeat for the desired number of repetitions

Purpose

To work the abdominal muscles at the front of the trunk (rectus abdominus)

Starting position and instructions

- Lie on your back with your knees bent and your feet firmly placed on the floor
- Tighten the abdominal muscles and pull them towards the back bone
- Maintain this fixed position of the abdominals throughout the movement
- Hold the kettlebells, so that the barrel rests in the crease of the shoulder
- Contract the abdominal muscles to lift and curl the shoulders and chest upwards
- Lift as far as is comfortable, without moving the lower back
- Reverse the movement under control

Coaching points

- Initiate the movement by contracting the abdominals and lifting the shoulders
- Take care not to hollow the back, keep the back fixed
- Keep the neck relaxed, following the movement of the rest of the spine
- Control the movement upwards and downwards
- If the hands are placed at the side of the head do not pull on the head

Progressions/adaptations/variations

- Vary the speed (2 counts up, 2 counts down or 3 counts up and 1 count down and vice versa)
- Increase repetitions to increase muscular endurance
- Use heavier kettlebells to increase resistance and challenge muscular strength
- A variation can be to lift the legs in the air and cross the ankles (a crunch)

Exercise 13.17	Twisting sit-ups/curl-ups with kettlebells

Purpose

To work the abdominal muscles at the side of the trunk (obliques). It will also work the muscles at the front of the trunk (rectus abdominus)

Starting position and instructions

- Lie on your back with your knees bent and your feet firmly placed on the floor
- Tighten the abdominal muscles and pull them towards the back bone
- Maintain this fixed position of the abdominals throughout the movement
- Hold the kettlebells, so that the barrel rests in the crease of the shoulder
- Contract the abdominal muscles to lift and curl the shoulders and chest upwards twisting the body towards the opposite side. Lower down and repeat this twisting to the other side

Coaching points

- Initiate the movement by contracting the abdominals and lifting the shoulders
- Take care not to hollow the back, keep the back fixed
- Keep the neck relaxed, following the movement of the rest of the spine

- Control the movement upwards and downwards
- Lift only as far as is comfortable, and without lifting the lower back off the floor
- Reverse the movement under control

Progressions/adaptations/variations

- Vary the speed (2 counts up, 2 counts down or 3 counts up and 1 count down and vice versa)
- Increase repetitions to increase muscular endurance
- Use heavier kettlebells to increase resistance and challenge muscular strength
- A variation can be to lift the legs in the air and cross the ankles (a crunch)
- Cycling the legs can provide further variation, however care must be taken not to twist too far or take the legs too far out. Lowering the legs too far will place unnecessary stress on the lower back

| Exercise 13.18 | Tricep dips with kettlebells |

Start position   End position

Purpose

To work the muscles at the back of the upper arms (the triceps)

Starting position and instructions

- From a seated position with knees bent place the kettlebells at each side of the body

- Hold the handles of the kettlebells, knuckles face down and wrists in fixed alignment
- Keep the feet flat on the floor and knees bent
- The buttocks are lifted
- Bend and straighten the elbow to lift and lower the body weight
- Perform for the desired number of repetitions

Coaching points

- Raise and lower the body under control
- Keep the abdominals pulled in tight and take care not to hollow the back
- Ensure the elbows extend fully as the body is lifted up, but do not lock the elbows
- Ensure the body is fixed and only the elbows move
- Keep the shoulders relaxed away from ears
- Keep the scapular in a fixed position, sliding down towards buttocks
- If the legs are extended or raised on a bench, take care not to lock the knee joint

Progressions/adaptations/variations

- Extend the legs to increase the resistance lifted
- Raise the feet onto a bench to increase the range of motion achieved
- Perform the exercise at a slower pace so that the muscles have to contract and work for longer
- Vary the speed of the exercise (2 down, 2 up or 4 down, 4 up)
- Vary the speed to focus work on different muscle contraction phases (3 down, 1 up to work on eccentric phase or 1 down, 3 up to work on concentric phase)
- Perform through different ranges of motion i.e. lower range or upper range
- Combine with a knee extension to add variation

Exercise 13.19	**Press-ups with kettlebells**

Purpose

To work the muscles at the front of the chest and shoulder (pectorals and anterior deltoid) and the muscles at the back of the upper arm (triceps)

Starting position and instructions

- Assume a full press-up position, with kettlebells on floor
- Hold the kettlebells with knuckles facing down and wrist fixed
- Select a wide or narrow hand position by placing the hands either shoulder width and a half apart and level with the shoulders, or shoulder width apart
- Bend and straighten the elbows to lower and lift the body weight up and down
- Perform for the desired number of repetitions

Coaching points

- Keep the abdominals pulled in tight and the back straight
- Keep the whole of the spine and neck in line
- Ensure the elbows fully extend but do not lock
- Keep the body weight forward and over the shoulders to maximise the resistance

- If kneeling in the $^3/_4$ position take care not to rest on the knee caps
- Ensure the body lowers in one smooth movement and lifts in one smooth movement
- Maintain a right angle (90 degrees) at the elbow joint on the lowering phase
- Keep the elbow and wrist in alignment
- If performing with narrow grip, ensure the elbows move backwards to maintain alignment

Progressions/adaptations/variations

- Progress from box to $^3/_4$ to full position
- Progress from full position by raising feet on a bench
- Increase repetitions to improve endurance
- Vary the speed of the exercise (2 down, 2 up or 4 down, 4 up)
- See previous exercise for other ways of varying speed

Exercise 13.20	**Gluteal raise with kettlebells**

Purpose

To work the muscles of the buttocks (gluteus maximus)

Starting position and instructions

- Adopt a full press-up position, holding the kettlebells with knuckles facing down and wrist fixed
- Fix the body and tighten the abdominals
- Raise one leg to hip height
- Perform for the desired number of repetitions
- Perform with other leg

Coaching points

- Keep the pelvis neutral and abdominals pulled in tight
- Keep the whole of the spine and neck in line
- Ensure the elbows are fully extended but not locked
- Raise the leg under control
- Keep the hips and shoulders square
- Keep the neck lengthened and scapula pressing down towards the buttocks

Progressions/adaptations/variations

- Increase repetitions to improve endurance
- Vary the speed of the exercise (2 down, 2 up or 4 down, 4 up)
- See previous exercise for other ways of varying speed
- Perform with knee bent at right angle to decrease lever length and increase range of motion
- Perform with press-ups

Summary

The key points discussed in this chapter include:

- An introduction to the use of kettlebells
- An illustration and explanation of exercises using kettlebells.

MEDICINE BALL TRAINING

14

'Old school' equipment can be used to train for total body power and explosiveness.

Medicine ball training is one of the oldest forms of strength and conditioning training. Nearly 3000 years ago, wrestlers trained with sand-filled bladders and in Ancient Greece, the physician Hippocrates had medicine balls sewn out of animal skins and stuffed with sand for his patients to throw back and forth, to use for injury prevention and rehabilitation.

Medicine ball training may be a simple concept but the movements themselves can be very demanding. Drills can be sport-specific or designed for general fitness. The bonuses of training with medicine balls compared to conventional free weights is that you do not need to stop the force applied to the weight and that the core muscles are constantly involved in all actions.

Medicine ball workouts build core trunk strength and joint stability. Building core trunk strength is an important foundation from which to begin building overall muscle strength. Functional arm and leg strength originates from the strength of the core or trunk (abdominals and back).

Joint stability is also vitally important because no matter how strong the muscles are they are only as functionally strong as the joints, which direct the muscle movements. In other words, you are only as strong as the weakest link in the chain, and that is your joints. All too often, strong and otherwise well muscled athletes break down because of weak joints in the shoulder, elbow, hips or knees.

Medicine ball training provides weight resistance through all planes of movement (frontal, transversal and sagittal). You gain the strength and flexibility necessary for explosive power motions – from a powerful golf swing to a booming tennis serve.

Medicine ball exercises

Exercise 14.1	Standing Russian twist

Start position   End position

Purpose

To work the abdominal muscles with focus on obliques

Starting position and instructions

- Stand with feet hip width apart
- Hold the ball level with the tummy button arms' length away from the body behind your right hip
- Forcefully swing the ball forward and round to the left hip
- Reverse back in the opposite direction; continue for the desired number of repetitions

Coaching points

- Keep a slight bend at elbow
- Keep the abdominal core muscles tight, turn/rotate back foot to allow greater range of movement

Progressions/adaptations/variations

- Reduce or increase the weight and size of the medicine ball
- Reduce the range of movement
- Reduce the force of the swing from side to side
- Seated Russian twists

Exercise 14.2 — **Diagonal chop**

Start position / End position

Purpose

To improve the strength and power of the trunk area through rotation and extension

Starting position and instructions

- Stand holding medicine ball level with knees in a slight squat position
- Raise the ball diagonally up towards the opposite shoulder keeping the arms straight
- At the same time rotate the trunk and extend through the knees
- Rotate or turn the rear foot to increase range of movement
- Repeat for the desired number of repetitions, and then repeat on the other side

Coaching points

- Keep knees in line with toes and feet facing forwards
- Keep the abdominals pulled in tight and take care not to hollow the back

Progressions/adaptations/variations

- Increase or decrease the weight and size of the medicine ball
- Reduce the force and range of movement of the chop
- Single leg chops

Exercise 14.3	Medicine ball lunge

Start position   End position

Purpose

To improve the muscular fitness of the gluteals, quadriceps, shoulders, upper back and trunk stabilisers

Starting position and instructions

- Pick up medicine ball and hold it on a bent arm level with tummy button
- While lunging forward raise the ball up over the head until arms are straight
- Return the ball back to the starting position when regaining the standing position
- Repeat using alternate legs for the desired number of repetitions.

Coaching points

- Stand with feet hip width apart
- Step forward keeping front knee in line with foot
- Bend back knee until a 90 degree angle at front and rear foot
- Keeping knees off the floor
- Keep a slight bend in the elbows
- Keep the abdominals pulled in tight and take care not to hollow the back

Progressions/adaptations/variations

- Increase or decrease the size and weight of the ball
- Decrease the range of movement of the lunge
- Hold medicine ball out in front on straight arms, when lunging forward rotate the trunk to the opposite side to the leading leg. Reverse ball to centre when stepping out of the lunge, repeat to either side

Exercise 14.4	Medicine ball slams

Start position   **End position**

Purpose

To improve the body's ability to generate strength and power, improve throwing ability and improve trunk stability

Starting position and instructions.

- Stand tall with feet over a hip width apart, holding the medicine ball in two hands, on a slightly bent arm, above and just behind the head
- Forcefully throw, pulling medicine ball down towards the ground with your mid-section
- Keep arms straight for as long as possible
- Catch the ball as it rebounds from the ground
- Repeat for desired number of repetitions

Coaching points

- Keep the abdominals pulled in tight
- Keep a slight bend in the knees
- Keep a slight bend in the elbows

Progressions/adaptations/variations

- Increase or decrease the size and weight of the medicine ball
- Reduce force of downward throw
- Do not catch on the rebound
- Use arms only
- Same exercise but with only one arm

Exercise 14.5   Figure of eights

Purpose

Increase functional core stability and help improve dynamic flexibility

Starting position and instructions

- Stand tall, feet wider than hip width apart
- Hold a medicine ball on extended arms over your right shoulder
- Bring the ball down in a continuous movement in front of you as if you are chopping wood, the ball ending towards your left foot
- Continue up toward your left shoulder then diagonally down towards your right foot

- Moving the medicine ball through figure of eight patterns
- Continue for the desired number of repetitions

Coaching points

- Throughout the action ensure that the abdominal region is held tight
- Keep a slight bend in the elbows

Progressions/adaptations/variations

- Increase the size and weight of the medicine ball
- Reduce the size of the figure of eight patterns
- Slow the action down

Exercise 14.6	One step wall throw

Purpose

Increase the upper body's ability to generate strength and power while maintaining trunk stability, this will also help improve throwing ability

Starting position and instructions

- Stand tall feet hip width apart
- Hold the medicine ball level with tummy button
- Lift and pull medicine ball up and behind the head with bent arms elbows facing slightly outwards
- Step forward with one leg and forcefully throw the medicine ball forwards and downwards, shifting weight forward onto the leading leg and bending at the hips
- Aim about 1–2 feet above the floor

Coaching points

- Keep knees in line with feet keeping a slight bend in the knee. Ensure that the abdominal muscles are pulled tight
- Keep a slight bend in the elbows

Progressions/adaptations/variations

- Increase the size and weight of the medicine ball
- Increase or decrease the throwing distance from the wall
- Without the step forward (Wall throw)

Exercise 14.9 | Side throws

Purpose

To improve strength and power of the trunk through rotation and to improve throwing ability

Starting position and instructions

- Stand sideways (90 degrees) to a wall or netting
- Rotate the trunk so that the medicine ball is behind the rear hip
- Support most of the body weight on the rear leg
- Rotate the trunk pushing the medicine ball forwards from hip height
- Turn through the balls of the feet transferring the weight to the forward leg as the ball is let go

- Catch the medicine ball on its rebound; continue for the desired number of repetitions

Coaching points

- Keep the abdominals pulled in tight and take care not to hollow the back
- Keep a slight bend in the knees
- Keep knees inline with feet

Progressions/Adaptations/Variations

- Increase or decrease the size and weight of the medicine ball
- Increase or decrease the rebound distance

Exercise 14.10	One leg squat

Purpose

To improve balance strength and power of the legs and to improve trunk stability

Starting position and instructions

- Stand tall on one leg holding the medicine ball at arms-length in front of you and level with the chest
- Squat down on one leg
- Keep the medicine ball held out in front, free leg held out in front also. Repeat the squatting action for the desired number of repetitions on each leg

Coaching points

- Keep a slight bend in the elbow
- Keep the knee in line with the feet
- Keep the abdominals pulled in tight and take care not to hollow the back
- Keep bottom above knees and below shoulders at lowest point of the squat

Progressions/adaptations/variations

- Increase or decrease the size and weight of the medicine ball
- Decrease the range of movement of the squat

- Ball can be held close to the body and pushed out and back only when squatting
- Free leg can be held to the rear or the side

Exercise 14.11	Toe touch with medicine ball

Purpose

Improve abdominal strength and power
Improve strength and endurance of the trunk

Starting position and instructions

- Lie supine on a mat head supported
- Legs should be straight and held in the air by bending at the hips

- Arms should be straight above the head holding a medicine ball facing the ceiling
- Lift the head and shoulders off the floor taking the medicine ball towards the feet (V sit action) by contracting the rectus abdominus
- Return under control to the starting position
- Continue for the desired number of repetitions

Coaching points

- Keep a gap between your chin and your chest
- Keep a slight bend in the elbows
- Keep the abdominals and pelvic floor muscles engaged throughout the action

Progressions/adaptations/variations

- Increase or decrease the size and weight of the medicine ball
- Keep feet flat on the floor and just concentrate on the upper body
- Release the ball into the air and catch on the way back down to the floor

Exercise 14.12	Seated Russian twist

Purpose

Improve the strength and power in the trunk through rotation
Improve trunk stability

Starting position and instructions

- Sit on your bottom with a slight lean back keeping head in its neutral position
- Knees should be bent and feet off the floor
- Holding a medicine ball, rotate the trunk so that the ball touches the floor on either side
- Repeat for the desired number of repetitions

Coaching points

- Keep a gap between the chin and the chest
- Keep the abdominals and pelvic floor muscles engaged throughout the action)

Progressions/adaptations/variations

- Increase or decrease the size or weight of the medicine ball
- Keep feet in contact with the floor
- Standing Russian twist

Exercise 14.13	Medicine ball press-ups

Purpose

Improve the strength and power of the pectorals, triceps and anterior deltoids
Improve trunk stability

Starting position and instructions

- Take up a prone front support position aligning ankle, knees, hips, shoulder and ear (bridge)
- Place the medicine ball under one hand to perform a press-up
- Do press-up with one hand on the floor next to the medicine ball pushing hard enough to land with the hand on top of the ball ready for the next rep
- Repeat for each rep done
- Continue for the desired number of repetitions for each side

Coaching points

- Keep a slight bend in the elbow at the end of the range of movement
- Keep the abdominals and pelvic floor muscles engaged throughout the action
- Maintain alignment with ankle, knees, hips, shoulders and ears

Progressions/adaptations/variations

- Decrease or increase the size of the ball
- Roll the ball across to the opposite hand between repetitions
- Use two medicine balls and keep a hand on each ball as you perform press-up

Exercise 14.14 Knee throw to push up

Purpose

Improve body's ability to generate strength and power in the upper body
Improve trunk stability
Improve throwing ability

Starting position and instructions

- Kneel on a mat
- Hold a medicine ball into the chest
- Throw the ball forward forcefully extending through the arms
- Follow the medicine ball catching yourself in a press-up position
- Complete the press-up, then repeat from the start
- Continue for the desired number of repetitions

Coaching points

- Keep hips in line with knees
- Keep the abdominals and pelvic floor muscles engaged throughout the action
- Keep a slight bend in the elbows at the end of the range of movement
- Maintain alignment with ankles, knees, hips, shoulders and ears
- Keep a gap between chin and chest

Progressions/adaptations/variations

- Increase or decrease the size and weight of the medicine ball
- Use a partner to return the medicine ball after the press-up has been performed which means exploding back to the kneeling position to receive it
- Use an overhead throw

Summary

The key points discussed in this chapter include:

- An introduction to the use of medicine balls
- An illustration and explanation of exercises using medicine balls.

SANDBAG TRAINING

15

A sandbag is an easy piece of equipment to make, it is nothing more than a heavy-duty, reinforced sack filled with builders' sand such as the sandbags that are used to stop floodwater. Sandbags can be made from a number of sources. These include:

- Ex-service kit bags (available from army surplus stores) – these are really effective for making heaver bags
- Sacks
- Even the humble pillowcase can be doubled up (more than one) to make a lighter sandbag

Sandbag training is very versatile, easy to learn and extremely challenging to the whole body. Sandbags by nature can be bulky and not always symmetrical, the sand inside can move thereby making the bag unbalanced. This provides more of a challenge during lifting, and when moving in various directions/movement planes.

Exercises using sandbags can on the whole mimic those performed using normal gym kit, such as barbells and dumbbells (Sandbags are a fraction of the cost of this equipment). However, there are certain specific exercises that can only be performed using a sandbag.

Sandbags can be very effective for functional strength and endurance training. Functional training is often defined as an exercise that copies a specific task (the specificity principle discussed in chapter 1. Functional activities may include an everyday task like loading the boot of the car when going on holiday or a specific sporting activity. For example, to become a better golfer, training should focus on exercises that mimic a golf swing etc. Sandbags offer an effective and inexpensive mode for improving functional strength and endurance.

A sandbag will challenge multiple muscles; not only will the targeted muscle groups gain benefit, but greater benefits to stabiliser muscles will also be achieved, in particular, trunk stability and grip strength. With a sandbag the first challenge is to pick this unbalanced bag up off the floor and get it into position, unlike a weights bar that can be taken off a rack, when already in the start position such as bench press and squatting. This challenges basic strength, coordination and technique. In addition, placing a sandbag on one side of the body means that the core muscles will have to work very hard to balance the body and keep the posture upright. One postural consideration would be to change the carrying side of the sand bag.

In the famous book *Dinosaur Training*, Brooks Kubik states: 'You feel sore as you do because the bags (sandbags) worked your body in ways you could not approach with a barbell alone. You got into the muscle areas you normally don't work. You worked the "heck" out of the stabilizers.' (Brooks D Kubik: 2006.)

With all these muscle groups working together a higher calorific expenditure can be expected during the workout. In addition, increased muscle mass, developed after long term training, will give increased calorific expenditure throughout the day. A further benefit of this type of training means that individuals can perform exercises for developing strength and endurance at home in their back garden or the garage.

Sandbag exercises

Exercise 15.1	Sandbag deadlift to modified snatch

Purpose

To improve muscular endurance and strength
To improve the body's ability to generate strength and power
To improve functional trunk stability and grip strength

Starting position and instructions

- Stand with feet a little wider than hip width apart, with a sandbag between the feet
- Bend at the hips and knees keeping knees in line with feet
- Keeping a straight and fixed back and tight abdominals reach down and grab hold of each end of the sandbag
- Leading with the shoulders and driving through the legs stand up
- At the same time keep the arms straight, a slight bend at the elbows and lift the sandbag above the head
- Reverse the action and lower the sandbag to the floor
- Repeat for 10–15 reps

Coaching points

- Keep knees in line with feet and keep bottom above knees and below shoulders at lowest point of squat
- Keep the abdominals and pelvic floor muscles engaged throughout the action
- Keep a slight bend in the knees at the top of the range of movement
- Keep slight bend in the elbows when fully extending them

Progressions/adaptations/variations

- Increase or decrease the size and weight of the bag used
- Split the action into two separate parts deadlift then snatch.

Exercise 15.2	Sandbag bear hug walking

Purpose

To improve strength, endurance and functional trunk stability

Starting position and instructions

- Deadlift the sandbag until you can hold it against the chest in a 'bear hug' type hold
- Walk a desired, challenging distance (4 x 20 metres?)
- Rest and repeat

Coaching points

- Keep knees in line with feet, bend at the hips and knees, keeping bottom above knees and below shoulders as in the deadlift
- Keep a slight bend in the knees at the top of the range of movement
- Walk with a heel through toe action
- Keep the abdominals and pelvic floor muscles engaged throughout the action

Progressions/adaptations/variations

- Increase or decrease the size and weight of the sandbag
- Increase or decrease the distance walked

Exercise 15.3 | Deadlift snatch 180 degree rotation

Purpose

Improve muscular strength and endurance and functional trunk stability
Improve grip strength and coordination

Starting position and instructions

- Grab the ends of the sandbag situated between the legs
- Deadlift the sandbag and lift up above the head keeping arms straight (with slight bend at elbow)
- Just before the bag reaches the top of the movement start to rotate the bag so that the front hand goes to the back and back hand to the front
- Lower the bag to the floor with the hands in this new position
- Repeat the movement for required number of repetitions

Coaching points

- Knees in line with feet, bend at the hips and knees keeping bottom above knees and below shoulders as in the deadlift
- Keep the abdominals and pelvic floor muscles engaged throughout the action

- Keep a slight bend in the elbows when arms are at full extension

Progressions/adaptations/variations

- Increase or decrease the weight (sand content) and size of the sandbag
- Deadlift and snatch only

Exercise 15.4	Headlock squat

Starting position and instructions

- Deadlift the sandbag and place under one arm in the form of a headlock. Bend at the hips and knees and squat down until thighs are level with the floor (with caution a deeper squat can be performed)
- Return to the standing position then repeat
- After the desired number of repetitions have been achieved change sides and repeat using the same amount of repetitions

Coaching points

- Keep the abdominals and pelvic floor muscles engaged throughout the action
- Knees in line with feet, bend at the hips and knees keeping bottom above knees and below shoulders as in the deadlift
- Keep heels on the floor

Progressions/adaptations/variations

- Increase or decrease the weight (sand content) and size of sandbag
- Lift sandbag from a bench

Purpose

Improve strength and endurance of the legs and functional trunk stability

Exercise 15.5	Half circle overhead snatch

Purpose

Improve strength and endurance and functional trunk stability

Improve grip strength

Starting position and instructions

- Place the sandbag on the floor outside one of your legs
- Reach down and grab hold of the sandbag, keeping legs slightly bent
- Extend through the legs and hips lifting the sandbag up and over the head, keeping arms straight (with slight bend at elbow)
- Place the sandbag back down on the opposite side
- Repeat this back and forth movement for the desired number of repetitions

Coaching points

- Keep the abdominals and pelvic floor muscles engaged throughout the action
- Keep hips facing forward
- Knees in line with feet

Progressions/adaptations/variations

- Increase or decrease the weight (sand content) and size of the sandbag
- Reduce the range of movement by having a bench on both sides using that instead of the floor

Exercise 15.6	Diagonal lift to shoulder

Starting position and instructions

- Reach down and grab hold of sandbag, which is on the floor by your left or right foot
- Pick up the sandbag lifting it across the body onto the opposite shoulder
- Reverse the movement and return sandbag to the floor
- Repeat for the desired number of repetitions
- Repeat exercise on the other side

Coaching points

- Keep the abdominals and pelvic floor muscles engaged throughout the action
- Keep a bend in the knees when picking up and lowering the sandbag
- Keep a slight bend in the elbow when arms are fully extended

Progressions/adaptations/variations

- Increase or decrease the weight (sand content) or size of the sandbag
- Reduce range of movement by elevating sandbag onto a bench as its starting point

Purpose

Improve strength and endurance and functional trunk stability
Improve grip strength and coordination

Exercise 15.7	Sandbag farmer's walk using sandbag

Purpose

Improve strength and endurance and functional trunk stability
Improve grip strength

Starting position and instructions

- Grab the end of the sandbag(s) with one hand and stand up
- Allow the sandbag(s) to hang down
- Walk for a given distances/time
- Repeat with other hand if only using one sandbag

Coaching points

- Heel through toe action, feet facing forwards and knees in line with hips and feet
- Keep the abdominals and pelvic floor muscles engaged throughout the action

Progressions/adaptations/variations

- Increase or decrease the weight (sand content) and size of the sandbag
- Use two sand bags
- Increase or decrease the distance walked prior to resting

Exercise 15.8   Sandbag Turkish getups

Purpose

Improve strength and endurance and functional trunk stability
Improve balance and coordination

Starting position and instructions

- Lie face up on the floor holding a sandbag on straight arms above your chest. Get into a sitting position with arms straight, keeping a slight bend in the elbows, above head
- Bring both legs underneath as if in a squat position keeping knees in line with feet and stand up
- Reverse the movement to lie down again
- Repeat for the desired number of repetitions

Coaching points

- Keep the abdominals and pelvic floor muscles engaged throughout the action
- Keep a slight bend in the elbows
- Keep knees in line with feet

Progressions/adaptations/variations

- Increase or decrease the weight (sand content) and size of the sandbag
- Start from a seated position
- Start from a kneeling position

Exercise 15.9 | Military press

Purpose

Improve strength and endurance and functional trunk stability

Improve grip strength

Starting position and instructions

- Deadlift and clean the sandbag to shoulder level
- Widen the foot stance to wider than hip width apart, slight bend in the knee
- Stand tall with good spinal alignment
- Press the sandbag over the head until the arms are straight (slight bend in elbows)
- Return to the start position, level with shoulders
- Repeat for the desired number of repetitions

Coaching points

- Knees in line with feet, bend at the hips and knees keeping bottom above knees and below shoulders as in the deadlift
- Keep the abdominals and pelvic floor muscles engaged throughout the action
- Keep a slight bend in the elbows when arms are at full extension

Progressions/adaptations/variations

- Increase or decrease the weight (sand content) and size of the sand bag
- Decrease the range of movement
- Cheat using a small squat action to gain momentum

Exercise 15.10 | Round back deadlift

Purpose

Improve strength and endurance and trunk stability

Improve balance

Starting position and instructions

- Squat down and grab hold of the sandbag, grapple it onto the chest as if holding in a bear hug
- Extend through the legs keeping knees in line with toes and stand up (the back will be rounded throughout this movement), ensure abdominal and pelvic floor muscles are held tight

- Return to the starting position and repeat the exercise for the desired number of repetitions

Coaching points

- Knees in line with feet, bend at the hips and knees keeping bottom above knees and below shoulders during the squat
- Keep the abdominals and pelvic floor muscles engaged throughout the action
- Keep a slight bend in the knees when legs are fully extended.

Progressions/adaptations/variations

- Increase or decrease the weight (sand content) and size of the sandbag
- Reduce the range of movement by elevating the sandbag onto a bench.

Exercise 15.11	Zercher walk

Purpose

Improve strength and endurance and functional trunk stability
Improve balance

Starting position and instructions

- Deadlift and grapple the sandbag up onto the chest, cradling with arms underneath in this position
- Walk forwards for the desired distance
- Rest then repeat for the desired number of repetitions

Coaching points

- Keep a heel through toe action with knees in line with hips and feet
- Spine should be extended and a neutral spine maintained (straight line between hips, shoulders and ears) with a gap between chin and chest

Progressions/adaptations/variations

- Increase or decrease the weight (sand content) and size of the sandbag
- Increase or decrease the distance to be walked before resting
- Start with sandbag elevated on a bench

Exercise 15.12	Single leg sandbag deadlift

Purpose

Improve strength and endurance and functional trunk stability
Improve balance

Starting position and instructions

- Deadlift the sandbag from the floor
- Hold the sandbag into the chest with arms cradled underneath
- Stand on one leg with a very slight bend at the knee joint
- Bend forward at the hips ensuring that the abdominal and pelvic floor muscles are held tight
- Continue to lean forward until you feel a comfortable tension in the hamstring muscle
- Return to the start position then repeat for the given number of repetitions then change legs

Coaching points

- Keep knee in line with foot
- Keep a slight bend in the knee
- Keep the abdominals and pelvic floor muscles engaged throughout the action

Progressions/adaptations/variations

- Increase or decrease the weight (sand content) and size of the sand bag

- Reduce the range of movement in the lean forward

Exercise 15.13	Backward sandbag drag

Purpose

Improve strength and endurance and functional trunk stability
Improve grip strength

Starting position and instructions

- Squat keeping knees in line with hips and feet
- Keep back in a good neutral position with hips, shoulders and ears aligned. Grab hold of one end of the sandbag
- Maintaining this squat position, walk backwards for a desired distance and rest
- Continue for the desired number of repetitions

Coaching points

- Keep the abdominals and pelvic floor muscles engaged throughout the action
- Keep knees in line with hips and feet
- Maintain a neutral spine throughout

Progressions/adaptations/variations

- Increase or decrease the weight (sand content) and size of sandbag

- Increase or decrease the distance the sandbag has to be dragged
- Change the direction of pulling to be laterally while maintaining a forward facing squat position

Sandbag circuit

Sandbags will be filled to three levels.

Large = Full sand bags
Medium = Three quarter full sandbags
Small = Half full sandbags

Twelve exercises are planned for the sandbag circuit. Each exercise can be performed for approximately one and a half minutes. Two complete circuits would take approximately 40 minutes.

Information on making and/or buying sandbags

For information on making a sand bag go to: http://www.crosstraining.com/sandbagconstructionkit

For information on buying a commercially produced sandbag, go to Powerbag© http://www.performt.com/about.html

Summary

The key points discussed in this chapter include:

- An introduction to the use of sand bags
- An illustration and explanation of exercises using sand bags
- An example of a circuit using sandbags.

Table 15.1	Sandbag circuit	
Exercise	Type sandbag	Starting position and instructions
Squat	Large	Stand with feet approximately shoulder-width apart with a sandbag on shoulders. Bend at the hips and knees with back straight keeping your heels on the ground. Squat until thighs are parallel to the ground and return to starting position.
Lunge	Large	Stand with feet approximately shoulder-width apart with a sandbag on shoulders. Lunge forward by stepping with left foot. Bend the back knee until the thigh of the front leg is parallel with the ground. Ensure the left knee is inline and behind the foot; ensure that the back remains straight. Return to the starting position and repeat action with the right leg.
Calf raise	Large	Stand with feet hip width apart and sandbag on shoulders. Rise on toes and balls of feet to full extension, then lower heels slowly back down. A board under the toes may be placed to increase range of motion.
Press-ups	Large, medium, small	Assume front support position. Use a spotter to ensure that sandbag remains on upper portion of back. Bend elbows keeping body straight until upper arm is parallel to the ground, then return to the starting position.
Modified bent-over row	Large, medium	Grab sandbag with both hands keeping head up, knees bent, with a slight lean forward at the hips. Bend elbows, pull sandbag in a straight motion to the lower portion of the chest and slowly return to the start position.
Military/shoulder press	Large, medium, small	Stand with feet shoulder-width apart. Clean the sandbag to chest level. Maintain a neutral spine. Keeping elbows out, press sandbag upwards until arms are straight with a slight bend in the elbow. Return to start position and repeat (exercise can be performed using one or two sandbags).

Table 15.1 Sandbag circuit (cont.)		
Exercise	Type sandbag	Starting position and instructions
Shoulder shrugs	Large, medium, small	Stand with feet approximately shoulder-width apart. Grab one sandbag in each hand allowing them to hang at arm's length. Maintain neutral spine. Raise shoulders (elevate) toward ears. Squeeze at the top position, then lower shoulders to start position.
Upright row	Large, medium	Stand with feet approximately shoulder-width apart. Grab one sandbag in each hand allowing them to hang at arm's length. Maintain neutral spine. Lift sandbags close to the body up towards chin while keeping elbows high. Lower slowly to start position.
Triceps extension	Large, medium, small	Stand with feet approximately shoulder-width apart. Place hands behind head with elbows close to your ears. Keep back straight. Spotter puts sandbag in hands. Raise sandbag to full extension while keeping elbows close to your ears and with a slight bend. Slowly return sandbag to start position.
Bicep curls	Large, medium, small	Stand with feet approximately shoulder-width apart. Grab sandbag with each hand and allow them to hang at arm's length. Maintain a neutral spine. Curl sandbag with one hand (palm up) to shoulder height. Lower sandbag slowly to start position. Repeat action with other arm (can be performed using one or two sandbags).
Deadlift	Large	Stand with feet approximately hip-width apart. Bend at hips and knees and assume a squatting position. Maintain neutral spine. Grab sandbag from a squatting position and stand up. Lower slowly to the starting position.
Curl-ups	Large, medium, small	On your back lie on the floor, knees bent, feet flat on the floor. Place sandbag on chest. Grab sandbag and raise body curling through the spine (30 degrees). Slowly lower back to the floor.

Adapted from http://www.globalsecurity.org/military/library/report/call/call_97–12_sustngc5.htm

SLEDGEHAMMER TRAINING

16

Sledgehammers are not normally regarded as a piece of equipment used for physical fitness conditioning. They are most likely seen as a piece of equipment used for manual labour, as a means of demolition.

However, sledgehammer training, as a muscular conditioning regime, has been around for a long time. It is normally associated with hard core training regimes.

Sledgehammer training may not look pretty but is an inexpensive method for providing an extremely effective whole body workout as part of a fitness training regime.

Sledgehammers are not the type of equipment you are likely to find in a studio or gym equipment store, but they can easily be stored next to the body bars, stability discs and weights bars.

According to coach Jamie Hale (2007), the muscles that benefit from swinging the sledgehammer are the abdominals, superficial and deep upper and lower back muscles, spine and trunk rotators, lateral flexors, backside, thigh and calves (whole body). Some of the benefits from swinging a sledgehammer include:

- Rotational strength and power
- Core strength
- Wrist and forearm strength
- Grip strength
- Increased dynamic range of movement
- Coordination
- Mental toughness.

(J. Hale: 2007)

Sledgehammers can be purchased from most DIY stores throughout the country. They come in a variety of weights: 3kg, 4.5kg and 6kg with heavier sledgehammers also available. Having a variety of resistances available bodes well for progression but avoid the temptation to train too heavy too soon. Frequently, less resistance offers more gain, as speed is paramount in this type of training.

Sledgehammer training requires the use of a lorry or tractor tyre (as the striking object) to absorb the impact of each sledgehammer swing. This type of training can be provided both indoors and outdoors, provided the appropriate health and safety checks have been made.

When you strike the tyre, you can expect the hammer to rebound slightly upon impact. The rebounding nature of the tyre will enhance wrist stability and strengthen the forearms and grip.

The following sledgehammer swings are the most commonly used conditioning exercises, diagonal swings, vertical swings and horizontal swings.

Sledgehammer exercises

Exercise 16.1	**Diagonal swing**

Purpose

Improve strength and power of the swing
Improve functional trunk stability
Improve grip strength and coordination

Starting position and instructions

- Stand a suitable distance from the rubber tyre (test with a slow motion action to check sledgehammer does not miss the tyre)
- Start by using the preferred side, holding the shaft near the bottom of the handle, lift the sledgehammer up and over the right shoulder (preferred side)
- Slide the top hand up towards the head of the hammer
- When swinging the hammer down and across the body, the hand at the bottom of the handle leads and guides the sledgehammer and should not move

- The top hand slides down the handle to meet the other hand. Grip tight. During the downward swing both the knees and hips will bend
- Return to the start position and repeat for the desired number of repetitions or desired time
- Repeat exercise on the non-preferred side

Coaching points

- Keep knees in line with feet and really pull the abdominals and pelvic floor tight

Progressions/adaptations/variations

- Increase or decrease the weight of the sledgehammer
- Increase or decrease the speed of the movement
- Alter the range of movement until comfortable with the technique

Exercise 16.2 — Vertical swing

Purpose

Improve strength and power of the swing
Improve functional trunk stability
Improve grip strength and coordination

Starting position and instructions

* Start by standing a suitable distance from the rubber tyre; again test this with the sledge-hammer resting on the impact site first
* Stand with feet facing forward, shoulder width apart
* Place the sledgehammer on the tyre in front of you, grip it with both hands at the end of the handle, and slowly raise it above and behind your head, keeping the knees in line with the feet, pushing hips forward
* Squeeze the handle hard so you have full control of the sledgehammer throughout the movement (alternate your hand positions every so many repetitions to make sure the mechanics of the swing are evenly distributed throughout your body)
* As you pull the sledgehammer forwards and downwards lead the movement by pushing the bottom backwards and elbows forwards, extend through the elbows keeping them slightly bent

* Power the sledgehammer to hit the tyre
* Repeat for desired number of repetitions

Coaching points

* Keep the abdominals and pelvic floor muscles really tight
* Keep knees in line with feet with feet facing forward
* Keep a slight bend in the knees when fully extended
* Keep a slight bend in the elbows when arms are fully extended

Progressions/adaptations/variations

* Increase or decrease the weight of the sledge-hammer
* Decrease or increase the speed
* Alter the range of movement until comfortable with the technique

Exercise 16.3	Horizontal swing

Purpose

Improve strength and power of the swing
Improve functional trunk stability
Improve grip strength and coordination

Starting position and instructions

- Lean and secure the tyre against a solid wall
- Stand at right angles to the tyre
- Feet should be shoulder width apart with the foot that is closest to the tire about half a meter away
- Holding the sledgehammer in the same way as for the diagonal swing, lift to the opposite side to the tyre so that the sledgehammer head is level with your head
- Start the swing without moving your feet by rotating through the trunk, bring both hands together at the bottom of the handle and grip tight
- Increase hip rotation by rotating through the

ball of the rear foot. Continue the swing while trying to hit the tyre at a level somewhere between your waist and chest areas, preferably just above your tummy button. Continue to swing for the desired number of repetitions or desired time then change sides and repeat

Coaching points

- Keep knees in line with feet
- Rotate on the balls of the feet
- Keep a slight bend in the elbow when arms are fully extended
- Keep the abdominals and pelvic floor muscles really tight

Progressions/adaptations/variations

- Increase or decrease the weight of the sledgehammer
- Decrease or increase the speed

Exercise 16.4	Golf swing

Purpose

Improve strength and power of the swing
Improve functional trunk stability
Improve grip strength and coordination

Starting position and instructions

- Stand about 30 to 50 centimetres away from the tyre, standing sideways to it
- Place your feet about shoulder width apart, feet facing forwards
- Grip the handle as you would a golf club by placing your right hand below your left on the handle, grip tight

- Raise the sledgehammer as far to the right as you can while keeping the wrist straight and firm
- During this lifting phase rotate the trunk and, to increase the rotation at the hips, rotate the leading foot
- At the top of the swing, forcefully yet under control, reverse the movement until you hit the tyre
- Repeat for the desired number of repetitions then change sides and hand positions

Coaching points

• Keep knees in line with feet
• Rotate on the ball of the foot
• Keep a slight bend in the knees when fully extended
• Keep a slight bend in the elbows when fully extended
• Keep the abdominals and pelvic floor muscles really tight

Progressions/adaptations/variations

• Increase or decrease the weight of the sledge-hammer
• Decrease or increase the speed
• Remove the tyre and add the swing follow through

Sledgehammer circuit

The following circuit can be performed in any order and at any number of repetitions up to those stated. Make sure that both sides are exercised and leading hands are changed round.

Progressions/adaptations/variations

• Increase or decrease the weight and size of the sledgehammer
• Decrease or increase the speed of movement
• Decrease the range of movement used
• Change the order of swings

Summary

The key points discussed in this chapter include:

• An introduction to the use of sledgehammers
• An illustration and explanation of exercises using sledgehammers
• An example of a circuit using sledgehammers.

Table 16.1	Sledgehammer circuit	
Type of exercise	Number of repetitions	Sledgehammer actions
Diagonal swing	25 reps each side	Starting with the sledgehammer above and behind one shoulder, swing down and diagonally across the body.
Vertical swing	25 reps	Start holding sledgehammer at the bottom of the handle, holding hammer above and behind the head, swing forward and down.
Shovelling (clearing a path)	25 reps each side	Stand holding the sledgehammer as if it's a garden spade, shovel a path through the snow throwing it behind you.
Triceps extension	25 reps	Place sledgehammer behind you holding the handle at the end, extend through the arms (as if scratching your back).
Copping wood	25 reps each side	Swing the sledgehammer as if chopping a tree trunk; shorter punching chopping strokes can also be used.
Toe lifts	25 reps each foot	Place the head of the sledgehammer on your foot, lift the toes off the floor supporting the handle with your hands.
Churning butter	25 reps	Hold the sledgehammer in front of your head, hammer down, lift sledgehammer up and down above the head.
Squat jumps	25 reps	Hold the sledgehammer in front of you, grabbing handle halfway up on bent arms, squat and explode up off the floor.
Drive in the fence post	25 reps each side	Hold the sledgehammer over and behind one shoulder, swing hammer down as if making contact with a fence post about waist height.

RUBBER TYRE TRAINING

17

Rubber tyres offer an excellent, effective and inexpensive method for advanced strength and conditioning circuits. Tyres, such as those used on tractors, lorries or buses can weigh anything between 25kg to 175kg. They can usually be acquired from specialist tyre fitting companies at no cost – most companies are happy to give them away, rather than paying to have them disposed of elsewhere!

Conditioning exercises using tyres usually involve flipping or dragging the tyres to provide a training effect. These activities are often associated as 'strongman' activities, but they also have other training uses. Adventurers have been seen dragging a rubber tyre around behind them in preparation for walking assaults on the North and South Poles. Conditioning coaches have also seen the value of tyre training, by including tyre dragging and flipping into some athletes' conditioning programmes, as a method of building strength and speed of leg drive. Tyre flipping is another activity that can be performed using a tyre and is becoming more in vogue with coaches and personal trainers.

Both flipping and dragging activities offer a way of developing the posterior rear muscle chain, including the calves, hamstrings, gluts and lower back muscles, grip and core strength. As a cautionary note – both these activities are extremely hard work.

Rubber tyre exercises

Exercise 17.1	Tyre flipping

Purpose

Improve strength and power drive through the legs

Improve functional trunk stability
Improve coordination
Improve balance

Starting position and instructions

- Start with the tyre lying on its side in front of you
- Take a four-point stance; grab the tyre, fingers underneath, resting the chest against the tyre
- Lift the tyre by extending through the legs, keeping knees in line with feet and forcefully pushing forwards and upwards (try to avoid performing a straight deadlift); maintain a straight long spine, keep the abdominal and pelvic floor muscles pulled in
- As the tyre lifts high enough change the hand-grip to a pushing action (palms on tyre), continue to push the tyre until it flips
- Repeat the action for the required distance, time or number of flips

Coaching points

- Keep knees in line with hips and feet
- Maintain a neutral spine
- Keep abdominal and pelvic floor muscles pulled tight
- Keep a slight bend in the knees and elbows when they are fully extended

Progressions/adaptations/variations

- Increase or decrease the size and weight of the tyre
- Decrease or increase the distance/time/number of flips used
- Start with the tyre resting off the floor, reducing the range of movement
- When tyre has been flipped, perform a double-footed jump forwards in and out of the middle of the tyre, then back to start position

| Exercise 17.2 | Tyre dragging |

Purpose

Improve strength and power drive through the legs
Improve functional stability
Improve balance

Starting position and instructions

- Start by selecting an appropriate sized tyre for your fitness level, two can be tied together, one on top of the other, to increase the size and weight
- Attach a reasonable length of rope to the tyre. This can be done in several ways: by tying

a rope strop around the tyre and connecting the pulling rope via a carabiner clip or by drilling two holes through the centre of the tyre and attaching a U-bolt remembering to have a backing plate to stop the bolts being pulled through, again the pulling rope can be attached to this bolt

• Attach the other end of the rope to a manufactured harness – a weights belt or a tool belt. It can even be attached to a strong backpack ensuring the waist strap is in use

• Some type of padding may be required between the belt and body. You're now ready to pull or drag

Coaching points

The tyre can be dragged or pulled in several ways – forwards, sideways, and backwards. It can even be pulled hand over hand towards you. Whichever technique is used, make sure the following are maintained throughout:

• Correct joint alignment of hips, knees and feet

• Correct posture and a strong core by pulling abdominal and pelvic floor muscles tight

Progressions/adaptations/variations

• Increase or decrease the size and weight of the tyre

• Decrease or increase the reps, distance or time used

• Use a weighted sled or similar

• Ever thought of using a SUV or 4x4 to pull? Possibly the ultimate pull!

Table 17.1	Tyre circuit	
Type of exercise	Distance/ repetitions	Tyre action
Stationary pull	20 metres	Stand the marked distance from the tyre (attached to a rope) grab hold of rope and use hand over hand action to pull the rope/tyre to the set distance.
Forward pulling	20 metres	Stand with your back to the tyre with rope attached to rear via a harness or belt. Walk/run and pull the tyre for the marked distance.
Tyre flipping	5–10 metres	Start in four point (hands and knees) stance, push and lift tyre until it flips, then two footed jump in and out of tyre and run around to start position and repeat flip and jump for required distance.
Backward pulling	20 metres	Stand facing the tyre with rope attached to the front of harness/belt. Lean back and pull tyre for set distance.
Tyre flipping	5–10 metres	Start in four point (hands and knees) stance, push and lift tyre until it flips, then two footed jump in and out of tyre and run around to start position and repeat flip and jump for required distance.

a. Forward pull: lean slightly forward from the ankles and drive through the legs, pushing off the balls of the feet, use a powerful arm swing to help with balance and drive

b. Sideways pull: stand side on to the tyre, slight lean from the ankles away from the tyre. Start walking sideways by crossing legs alternating in front and behind the stationary leg (in a grapevine fashion)

c. Backward pull: stand facing the tyre, in a slightly seated position. Start to walk backwards, as you take the strain push forwards with the hips while maintaining a strong neutral spine

d. Stationary pull: stand facing the tyre in a strong upright position with a slight bend at the hips and knees. Grabbing hold of the rope pull hand over hand, keeping the elbows close to the body, until tyre is covered the pulled distance

Perform the circuit in a square. Other activities can be included as transitions between each tyre exercise if desired, but the tyre circuit is usually hard enough on its own.

Aim to complete three rounds of the circuit. Time the total time taken to use as a target to beat in the future. Remember to modify resistance and exercises to suit current fitness level.

Summary

The key points discussed in this chapter include:

- An introduction to the use of tyres
- An illustration and explanation of exercises using tyres
- An example circuit using tyres.

ROPE CLIMBING

18

Rope climbing is very much an old school training method. School PE lessons through the 1960s and 1970s would almost certainly have included some kind of rope climbing activity with ropes connected to the high roof beams. Rope climbing is/was possibly our first introduction to body weight strength training! It involved lifting the body weight off the floor and generally holding it suspended above the ground for a period of time.

Rope climbing was usually either loved or loathed by pupils. The heavier the pupil, the harder it was! Today, due to health and safety restrictions, rope climbing has fallen from the PE curriculum which, in many ways, is a shame. The only place it tends to be found is in the military, where rope climbing is a normal activity, used to improve leg, back, arm and grip strength, while also testing coordination and courage.

A three-strand 6.4cm diameter manila rope is a good-sized rope to use, however any reasonable diameter (natural fibre) rope will suffice. It is important to ensure that the top end of the rope is really well secured to an appropriate anchor point, and tested to hold the appropriate weight of the individuals who will be climbing them!

Exercise 18.1	Assisted pull-ups

Climbing techniques

Purpose

Increase arm and upper body strength
Grip strength

Starting position and instructions

- Stand behind the rope, reach up and grasp the rope with both hands, one higher than the other, while keeping both feet on the floor
- As you pull on the rope with your arms, extend through the ankles. Continue to pull on the arms until they are level with the chest
- Hold then lower under control
- Repeat for the desired number of repetitions

Coaching points

- Keep a slight bend in the elbows when arms are fully extended
- Keep abdominals and pelvic floor pulled in tight

Progressions/adaptations/variations

- Increase or decrease the height of the hands when pulling up
- Reduce the amount of extension through the legs
- Increase or decrease the speed

Exercise 18.2	Rope climb

Purpose

Arm and upper body strength
Grip strength

Starting position and instructions

- Stand behind the rope, reach up as high as you can and grasp the rope with both hands.
- Pull on the arms and lift the legs up towards the chest, allowing the rope to run down the inner thigh then across the shin of the right or left leg, grabbing the rope with both knees and the outside of the feet (do not stand on the rope)
- When the knees are as high as you can get them, lie back on straight arms, grip with the knees and feet and reach up with each hand to increase height climbed (a shift)
- Return to the ground hand under hand allowing the rope to slide through the legs
- Repeat for the desired number of shifts

Coaching points

- Keep a slight bend in the elbows when arms are fully extended
- Keep abdominals and pelvic floor muscles pulled in tight
- Maintain a neutral spine

Progressions/adaptations/variations

- Increase or decrease the number of shifts (height) climbed
- Increase or decrease the hip and knee bend when climbing

Exercise 18.3	Seated climb arms only

Purpose

Arm and upper body strength
Grip strength

Starting position and instructions

- Sit behind the rope with legs together out in front (pike position)
- Reach up the rope with the hands as high as you can

- Pulling on the hands, climbing hand over hand, keeping arms bent, pull your body to standing
- Lower under control, hand under hand, back to the seated position

Coaching points

- Keep a slight bend in the elbows when arms are fully extended
- Keep abdominals and pelvic floor pulled in tight

Progressions/adaptations/variations

- Start with legs bent rather than straight
- Reduce the height between sitting and standing to be climbed
- When at desired height perform a number of pull ups

Exercise 18.4	Arms only climb

Purpose

Increased arm strength
Grip strength

Starting position and instructions

- Standing behind the rope, reach up with two hands keeping the arms bent. Hand over hand climb the rope without the use of the legs or feet, until the desired height has been reached
- Lower hand under hand back down to the ground
- Repeat for the desired number of climbs

Coaching points

- Keep a slight bend in the elbows when arms are fully extended
- Keep abdominals and pelvic floor muscles pulled in tight

Progressions/adaptations/variations

- Increase or decrease the height to be climbed
- Decrease the time it takes to reach desired height
- Climb with legs trailing or in a pike position
- Climb by using the momentum of the legs chopping down

Exercise 18.5 — Climbing using two ropes

Purpose

Increased arm strength and coordination
Grip strength

Starting position and instructions

- Climb up two ropes hanging side-by-side without using your legs
- Reach up and place one hand on one rope and the other hand on the other rope
- Lean towards one of the ropes and quickly release and re-grasp the opposite rope several centimetres higher
- Then lean towards the other rope and simul-

taneously pull with that hand while quickly releasing and re-grasping the rope with the opposite hand several centimetres higher than the hand on the opposite rope
- You can also do this technique with your feet dangling or in L-seat position.

Coaching points

- Keep a slight bend in the elbows when arms are fully extended
- Keep abdominals and pelvic floor muscles pulled in tight
- Maintain a neutral spine

Progressions/adaptations/variations

- Increase or decrease the height to be climbed
- Use a spotter to help take the weight when holding on with one hand
- If more ropes are side-by-side, practice by traversing across the ropes with feet off the floor

Exercise 18.6	**Arm jump climbing**

Purpose

Arm and upper body strength
Grip strength

Starting position and instructions

- Stand behind and in the middle of the ropes, reach up and grab the two ropes at the same height

- Pull yourself up to approximately chest level quickly and powerfully to create momentum
- Let go of the ropes and quickly reach up to re-grab the ropes at a higher level, while your body is momentarily suspended in the air
- Height of the reach will depend upon how much momentum and power you can generate during the pull up
- Repeat until desired height has been reached.
- Lower to the floor under control, hand under hand using one or both of the ropes.

This is a tough exercise that requires a good level of strength and conditioning as well as coordination as it puts a big strain on elbows and shoulders. It should therefore only be done periodically to give a greater challenge. There is a risk of rope burns to the hands on the re-grab if the grip is not strong enough due to sliding down the rope.

Progressions/adaptations/variations

- Increase or decrease the height to be climbed
- Use small or large hand jumps
- Use only one rope

All of the above exercises should be performed with caution. Only use the relevant climb conducive to the fitness/skill level of the individual and progress over time to the others. A safety precaution would be to have a crash mat or mattress at the bottom of the rope, in case of falls.

Summary

The key points discussed in this chapter include:

- An introduction to the use of rope climbing
- An illustration and explanation of exercises using ropes.

TELEGRAPH POLE/GYM BENCH TRAINING

19

The average person would not associate telegraph poles or logs with fitness circuit training. However, with a little imagination and creativity, they are an amazing resource – you will be surprised at what can be done with them. The military services have used telegraph poles and logs for centuries as a means of improving muscular and cardio fitness, as well as for team building in the field.

An old-fashioned wooden gym bench normally found in school gyms or sports halls can be used as an alternative: this is what we have used in the photos for this section. Both need to be between 2 and 3 metres in length and generally would weigh a minimum of 20kg. Rope handles/strops can be fitted to the telegraph poles/logs for carrying while running, which increases their usefulness.

The following are examples of the type of exercises that can be performed with this type of equipment. Each exercise is performed with two, three or four people to a pole/bench.

Telegraph poles can be obtained from wood reclamation sites or by contacting your local forestry commission. As we say, we have used gym benches here as they are just as effective, but the exercises are generally known by their telegraph pole title.

Telegraph pole exercises

Exercise 19.1	The oblique deadlift to shoulder

Purpose

Increase the strength of the legs and upper body

Increase the strength of the oblique muscles
Increase coordination

Starting position and instructions

- Stand facing forward with the telegraph pole/log on the right hand side, on the floor
- Bend at the hips and knees leading with backside. Rotate the upper body to the right and take an over hand grip of the pole with the left hand and underhand grip with the right hand
- Leading with the shoulders lift the pole off the floor and up on to the shoulders, rotating the trunk to the left as you do the lift
- Return to the floor and repeat for the desired number of repetitions
- Repeat on the other side

Coaching points

- Bend at hips and knees pushing backside backwards and downwards
- Keep the knees in line with hips and feet
- Keep backside above the knees at the lowest point
- Brace the abdominal and pelvic floor muscles tightly

Progressions/adaptations/variations

- Increase or decrease the number of people lifting the telegraph pole
- Increase the size and weight of the telegraph pole

Exercise 19.2	The shoulder press

Purpose

Increase the strength of traps, deltoids and triceps

Starting position and instructions

- Stand holding the pole above the head with bent arms
- Leading with the tips of the fingers straighten the arms
- Leading with the elbows return the pole back to the start position
- Alternatively start with the pole on one shoulder, lift off the shoulder to full extension, then lower to the opposite side, repeat for the prescribed time/repetitions

Coaching points

- Keep a slight bend at the elbow at the top of the movement
- Keep abdominals and pelvic floor muscles pulled in tight
- Keep a slight bend in the knees
- Maintain a neutral spine position

Progressions/Adaptations/Variations

- If working in a pair one lifts while the other lowers.

Exercise 19.3	Squat

Starting position and instructions

- In pairs stand facing each other at either end of the pole
- The end of the pole should be resting in hands with fingers rapped around the pole just below the chin (the pole can be deadlifted into this position)
- Then, bending at the hips and knees, squat until thigh is in line with knees then stand up
- Repeat for prescribed number of repetitions or time.

Coaching points

- Keep knees in line with feet
- Keep abdominals and pelvic floor muscles pulled in tight
- Keep the backside in line with the knees at the lowest point of the squat

Progressions/adaptations/variations

- Perform the squat alternately to increase the intensity for each person
- Perform the squat while doing the shoulder press

Purpose

Increase the strength of the gluts and quads (legs)

Exercise 19.4	Lying bench press

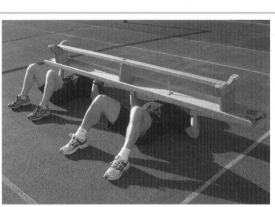

Purpose

Increase the strength of the pectorals, triceps and anterior deltoids

Starting position and instructions

- Lie on the floor with knees bent and feet flat on the floor hold or support the pole with straight arms a shoulder and a half width apart, in line with the chest
- Slowly, under control, lower the pole towards the chest until upper arm hits the floor
- Return press the pole back to the start position.

Coaching points

- Keep a slight bend in the elbow when arm is at full extension
- Keep abdominals and pelvic floor muscles pulled in tight
- Keep backside and feet in contact with the floor

Progressions/adaptations/variations

- Perform the squat alternatively to increase the intensity for each person
- Perform exercise on a slope with head high and feet low to work upper chest
- Perform exercise on a slope with head low and feet high to work lower chest

Exercise 19.5	Bent forward rowing

Purpose

Increase the strength of the latissimus dorsi and biceps

Starting position and instructions

- With one hand under and one hand over the pole, deadlift it and hold away from the body level with the knees
- Stand with feet hip width apart with a bend in the knee and a slight lean forward at the hip pulling in through the abdominal muscles
- Leading with the elbows pull the pole in towards the body, changing the grip position after several repetitions to balance the pulling arm

Coaching points

- Keep a slight bend in the elbows when arms are fully extended
- Keep abdominals and pelvic floor muscles pulled in tight
- Keep a bend in the knees throughout
- Maintain a neutral spine

Progressions/adaptations/variations

- Perform the squat alternatively to increase the intensity for each person

Exercise 19.6	Standing back extension

Coaching points

- Keep the knees slightly bent
- Keep knees in line with feet
- Keep the abdominals and pelvic floor muscles pulled in
- Maintain neutral spine

Progressions/adaptations/variations

- Perform this exercise while sitting on a bench or chair
- Perform this exercise without the pole as a normal back extension.

Exercise 19.7	Lying abdominal curl

Purpose

To increase the strength of the back muscles

Starting position and instructions

- Deadlift the pole and place both arms underneath and grip in close to the chest
- Take a slight bend at the knees keeping them in line with the feet, keeping the abdominal muscles tight, bend forward from the hips, about 30 degrees, and then back to the upright
- Perform this exercise slow and deliberately
- Repeat for the prescribed number of repetitions or time

Purpose

To increase the strength of the abdominal region (rectus abdominals)

Starting position and instructions

- Sit holding the pole in towards the chest and lower back down towards the floor
- Bend the knees with feet flat on the floor, keep the pole tight to the chest and curl up through the spine (about 30 degrees) keeping the abdominal muscles tight, then slowly lower down to the floor

- Repeat for the prescribed number of repetitions or time

Coaching points

- Keep abdominals and pelvic floor muscles pulled in tight
- Keep the chin off the chest
- Keep the backside and feet in contact with the floor

Progressions/adaptations/variations

- Perform the abdominal curl alternatively to increase the intensity for each person

Exercise 19.8	Suitcase carry

Purpose

To improve core stability
To improve grip strength
To improve aerobic conditioning

Starting position and instructions

- Holding the handles if fitted and carry as if carrying a suitcase
- Cover a predetermined route at a fast walking pace, change sides at a pre-measured half distance then continue

Coaching points

- Keep a heel through toe action
- Keep knees in line with hips and feet
- Maintain a neutral spine
- Keep abdominals and pelvic floor muscles pulled in tight

Progressions/adaptations/variations

- Changing the speed that you cover the course
- Include obstacles along the route that have to be negotiated

Log run

Running around a suitable course (1.5–5km) carrying the telegraph pole by its handles. The exercises listed previously can be included at various points around the circuit to make it more interesting.

Summary

The key points discussed in this chapter include:

- An introduction to the use of telegraph poles
- An illustration and explanation of exercises using telegraph poles.

LIST OF ABBREVIATIONS

ACSM	American College of Sports Medicine
ADP	adenosine di-phosphate
ATP	adenosine tri-phosphate
CHD	coronary heart disease
CNS	central nervous system
CP	creatine phosphate
CR	cardio respiratory
CRAC	contract, relax, antagonist contract
DOMS	delayed onset of muscle soreness
ETM	exercise to music
FG	fast glycolytic
FITT	frequency, intensity, time, type
FOG	fast oxidative glycolytic
GTO	Golgi tendon organ
HR	heart rate
HRR	heart rate reserve
KPH	kilometres per hour
MHR	maximum heart rate
MPH	miles per hour
OBLA	onset of blood lactate accumulation
PNF	proprioceptive neuromuscular facilitation
RM	repetition maximum
RPE	rating of perceive exertion
RPM	revolutions per minute
SAID	specific adaptation to imposed demand
SD	standard deviation
SPM	strokes per minute
THR	target heart rate
VO_2MAX	maximal oxygen uptake

REFERENCES AND RECOMMENDED FURTHER READING

ACSM (2000) 6th edition. *ACSM's Guidelines for Exercise Testing and Prescription*. USA: Lippincott, Williams and Williams.

ACSM (2006) 7th edition. *ACSM's Guidelines for Exercise Testing and Prescription*. USA: Lippincott, Williams and Williams.

Bean, A (2001) *The Complete Guide to Strength Training*. UK: A & C Black.

Brittenham, D and Brittenham, G (1997) *Stronger Abs and Backs*. USA: Human Kinetics.

Bodybuilding.com (author and date unknown). *Sledgehammer Training* http://www.body-building.com/sledgehammerGPP, accessed April–August 2007.

Bompa, T and Carrera, M (2005) *Periodization Training for Sports. Science-based strength and conditioning plans for 20 sports*. USA: Human Kinetics.

Bursztyn, P (1990) *Physiology for Sports People. A Serious User's Guide to the Body*. UK: Manchester University Press.

Chu, D Dr (1998) 2nd edition. *Jumping into Plyometrics*. USA: Human Kinetics.

Davis, D, Kimmet, T and Auty, M (1986) *Physical Education: Theory and Practice. Melbourne*. Australia: Macmillan Education.

Davis, R, Roscoe, J, Roscoe, D and Bull, R (2005) 5th edition. *Physical Education and the Study of Sport*. UK: Mosby.

Department of Health (2004). *At Least Five a Week. Evidence on the impact of physical activity and its relationship to health*. A report from the Chief Medical Officer. London: Department of Health.

Department of Health (2005). *Choosing Activity: A Physical Activity Action Plan*. London: Department of Health.

Doheny, Kathleen (2007) The 300 workout. From: www.webmd.com/fitness-exercise/features/the-*300-workout-can-you-handle-it?* accessed 15 November 2007.

Enamatt, R (date unknown) *Sledgehammer Training*. From: http://www.rosstraining.com, accessed April–August 2007.

Engles, R (2007) *Sledgehammer Workouts*. From: http://www.shovelglove.com, accessed April–August 2007.

Fleck, S and Kraemer, W (1987). *Designing Resistance Training Programmes*. Illinois. USA: Human Kinetics.

Gambetta, V and Odgers, S (1991) *The Complete Guide to Medicine Ball Training*. USA: Gambetta Sports Training Systems.

Global Security (author and date unknown). *Maintaining Physical Fitness*. From: http://www.globalsecurity.org/military/ttp, accessed August 2007.

Hale, J (date unknown) Sledgehammer. From: http://www.redwhiteandbluefitness.com, accessed July 2007.

Harris, J (1997) *Health related exercise in the national curriculum. Key stages 1 to 4*. UK. Loughborough University: Human Kinetics.

Hayes, F (1995) Fitness Programming *Guide Book*. Basingstoke. UK: SUMMIT Training and Education.

Hough, M (1998) *Counselling Skills and Theory*. UK: Hodder & Stoughton.

Kubik, B (August 2006) 5th edition. *Dinosaur Training – Lost Secrets of Strength and Development*. KY. USA: Brooks D Kubik.

Lawrence, D (2004a) 2nd edition. *The Complete Guide to Exercise in Water*. UK: A & C Black.

Lawrence, D (2004b) 2nd edition. *The Complete Guide to Exercise to Music*. UK: A & C Black.

Lawrence, D (2005) *The Complete Guide to Exercising away Stress*. UK: A & C Black.

Lawrence, D and Hope, B (2007) *The Complete Guide to Circuit Training*. UK: A & C Black.

Lawrence, D and Barnett, L (2006) *GP Referral Schemes*. UK: A & C Black.

Lee, R (2007) *Workoutpass*. North Palm Beach Florida USA. From: www.Workoutpass.com, accessed June 2007.

Mahler, M (2002–2007) Sandbag Training. From: http://www.mikemahler.com/sandbagtraining.mht, accessed June 2007.

McArdle, W. Katch, F and Katch, V (1991) 3rd edition. *Exercise Physiology. Energy, Nutrition and Human Performance*. USA: Lea & Febiger.

Petty, G (2004) 3rd edition. *Teaching Today*. UK: Nelson Thornes Ltd.

Reece, I and Walker, S (2003) *Teaching, Training and Learning*. Oxford. UK. Business Education Publishers.

Scholich, M (1999) *Circuit Training for all Sports. Methodology of Effective Fitness Training*. Toronto. Canada: Sports Book Publisher.

Sharkey, B (1990) 3rd edition. *Physiology of Fitness*. Illinois. USA: Human Kinetics Books.

Siff, Mel C (2003) 6th edition. *Supertraining*. USA: Supertraining Institute.

Siff, Mel C (2003) 6th edition. *Facts and Fallacies of Fitness*. USA: Mel Siff.

Tsatsouline, P (2002) *From Russia with Tough Love*. USA. Dragon Door Publications Webmd.com, accessed March 30 2007.

Yessis, M and Trubo, R (1987) *Secrets of Soviet Sports and Fitness Training*. USA: Arbour House Publishing.

INDEX